BFC 9323

BIOTIC HOMOGENIZATION

BIOTIC HOMOGENIZATION

Edited by
Julie L. Lockwood
University of California, Santa Cruz
Santa Cruz, California

and

Michael L. McKinney
University of Tennessee
Knoxville, Tennessee

Kluwer Academic / Plenum Publishers
New York, Boston, Dordrecht, London, Moscow

Library of Congress Cataloging-in-Publication Data

Biotic homogenization / edited by Michael L. McKinney and Julie L. Lockwood.
   p. cm.
  Includes bibliographical references (p.).
  ISBN 0-306-46542-6
   1. Biological diversity. 2. Ecology. I. McKinney, Michael L. II. Lockwood, Julie L.

QH541.15.B56 B66 2001
576.8--dc21

2001029060

QH
541
.15
.B56
B66
2001

ISBN 0-306-46542-6

©2001 Kluwer Academic/Plenum Publishers, New York
233 Spring Street, New York, N.Y. 10013

http://www.wkap.nl/

10 9 8 7 6 5 4 3 2 1

A C.I.P. record for this book is available from the Library of Congress.

All rights reserved

No part of this book may be reproduced, stored in a retrieval system, or transmitted in any form or by any means, electronic, mechanical, photocopying, microfilming, recording, or otherwise, without written permission from the Publisher

Printed in the United States of America

FOREWORD

Humans tend to mark historical time periods by words or phrases that capture the essence of significant events. At present, the transformational period is the millennium and Diversity the word. As we attempt to diagnose and struggle to solve the big problems of the world, Diversity underlies so much we are trying to change. Equal opportunity programs call for racial, gender, and multicultural diversity in society. Dieticians promote balanced food groups to avoid disease. Even athletic teams lacking a balanced attack falter in the heat of competition. Biologists, of course, have long recognized that the 'ecological theatre and evolutionary play' are driven by life's diversity. Today, this is at the heart of biodiversity studies: Describing, understanding, protecting, managing and conserving *Bio*diversity is the most critical challenge on the planet. Conservation biology, the 'crisis science', is now the designated research agenda for addressing how to preserve biodiversity in the face of human encroachments and impacts in all their forms.

With every important conceptual idea there is a flipside, an apostate that lends insight into a difficult problem. 'Biological Homogeneity' is that apostate to diversity. The idea is intuitive. Biological phenomena are becoming more uniform in time and space. Habitat destruction, overexploitation, invasions of exotic species and disease, all hallmarks of humans becoming the dominant species on earth, are winnowing away Earth's biodiversity. The underlying problem is that humans have taken over; wherever we have gone we affect the natural environment, create sprawling urban areas that destroy natural qualities, and promote cultural homogeneity through migration, travel, and communication. We live in a time where those who in the past shared little now increasingly come in direct contact spatially, ecologically, culturally, and politically. This so-called "population displosion" is raising the question of how human mixing will impose upon and potentially eliminate biodiversity.

We already know part of the answer, and it isn't pleasant. Biotic homogeneity is occurring at breakneck speed, undoubtedly the result of human population growth. Virtually every biological level – cellular, genetic, individual, population, species, higher taxonomic, ecological community, ecosystem – has been effected. There are depressingly obvious indicators just within our lifetimes including the demise of the Yellowstone ecosystem, marked extinctions of species such as the Columbian grebe and Caribbean monk seal, dramatic decline in songbirds, destruction of coastal waterways, and global climate change exceeding predictive models. The trend of human population growth must stop: 10 000 years ago there were 5.7 million people, today there are about 5.7 billion, a thousand-fold increase in human population size. At this rate, humans double in population once per millennium, a rate that would reach a catastrophic 5.7 trillion people in the next ten thousand years. With the planet 510 million square kilometers in size, that population would give each person less than 10 meters in which to reside. The next millennium must bring an abrupt reversal of the global population growth rate. As convincingly shown by the chapters in this book, we cannot permit population numbers to reach the point where the watchword for the next era of biodiversity becomes the word biomonotony.

John L. Gittleman
Charlottesville, VA

PREFACE

Biological homogenization is the dominant process shaping the future global biosphere. As global transportation becomes faster and more frequent, it is inevitable that biotic intermixing will increase. Unique local biotas will become extinct only to be replaced by already widespread biotas that can tolerate human activities. This process is affecting all aspects of our world: language, economies, and ecosystems alike. The ultimate outcome is the loss of uniqueness and the growth of uniformity. In this way, fast food restaurants exist in Moscow and Java Sparrows breed on Hawaii.

Biological homogenization qualifies as a global environmental catastrophe. The Earth has never witnessed such a broad and complete reorganization of species' distributions. Even the greatest homogenization episode of all, the unification of all the major continents into the supercontinent Pangea, cannot hold a candle to the forthcoming Homogecene. Humans not only disperse species to places where they could never reach on their own. We also homogenize the physical environment by constructing the same repetitious habitats, composed largely of concrete and simple monocultural floras, all over the planet. We challenge the reader to randomly place his or her finger on a map of the World and describe its modern biological attributes without producing a list of exotic species followed by a list of extinctions. If you so choose, you can ignore all domesticated animals and crops, as most of these will not be native to the region you have randomly identified. You can also ignore any extinctions that we know only through the discovery of fossils. The result will be the same. There are very few places that, if compared biologically to themselves in 1700, would be easily recognizable as the same place.

Some regions are far more modified than others and some taxa are doing far better from human activities than others. The goal of this volume is to identify regions and taxa that either suffer, or benefit, inordinately from this global reorganization process. What can we learn from past biological reorganizations? What patterns can we identify today that may lead to proactive conservation goals aimed at reducing homogenization? If we project current trends into the future, what will a *more* homogenized world look like? Asking and answering these questions are the conservation ecologist's way of dealing with, and hopefully providing answers to, the problem.

Perhaps the most troubling aspect of biological homogenization is how it could degrade the human perception of "nature". One might call this the boredom factor. Biological homogenization has left regions that were once known for their ecological beauty and uniqueness (e.g., Hawaii, African Rift Lakes, and even the Central Valley of California) as biological replicates. When replicates are produced, the natural world becomes a boring place. People begin to care less about the species they see, as more and more of the things they *do* see are 'as common as dirt'. Where once you had to travel the Atlantic to see the smart gray and black 'overcoat' of the male English sparrow, it is now possible to see this bird in almost any urban area of the world. Most people couldn't even tell you the name of this bird even though it is literally under their feet every day of their lives. Those of us that do know that this little gray and brown bird is *Passer domesticus* often consider it beneath our contempt as it thrives in cities, eats our garbage, and does nothing to enhance our life lists. Where before you had to make your way to the clear, fast streams of North America to see the exquisitely colored Rainbow Trout, now one only need drive as far as the local reservoir. Catching rainbow trout is no longer the fisherman's prize, it is the acceptable outcome of a day of casting a fishing line.

We love the things we understand and we value the things that are unique. Biologically unique experiences are now confined to National Parks, Biological Reserves, and other set-asides. The day to day world we inhabit is filled to exhaustion with the common, and thus the mundane. Biological homogenization creates more than an impaired global environment, it creates a biologically indifferent culture. In the end, our human appetite for variety will be cheated. To paraphrase the conclusion of James Kunstler's wonderful book on cultural homogenization appropriately entitled *The Geography of Nowhere*: if every place is the same, why go anywhere?

Given the incoming tidal wave of the global economy, attempts to completely stop biological homogenization seem futile indeed. Yet educated conservationists around the world must at least try to stem the tide. We have no choice, as conservationists have known for centuries, because to surrender the Earth to human needs alone is to surrender many of the most valuable treasures it has. With this reality in mind, we put forth this volume.

Julie Lockwood *Michael McKinney*
Santa Cruz, California Knoxville, Tennessee

Contents

Chapter 1. **Biotic Homogenization: A Sequential and Selective Process**   1
Michael L. Mckinney and Julie L. Lockwood

Chapter 2. **Biotic Homogenization: Lessons from the Past**   19
Kaustuv Roy and Jeffrey S. Kauffman

Chapter 3. **Birds and Butterflies Along Urban Gradients in Two Ecoregions of the United States: Is Urbanization Creating a Homogeneous Fauna?**   33
Robert B. Blair

Chapter 4. **Rarity and Phylogeny in Birds**   57
Thomas J. Webb, Melanie Kershaw and Kevin J. Gaston

Chapter 5. **Hybridization between Native and Alien Plants and Its Consequences**   81
Curtis C. Daehler and Debbie A. Carino

Chapter 6. **Taxonomic Selectivity in Surviving Introduced Insects in the United States**   103
Diego P. Vázquez and Daniel Simberloff

Chapter 7. **Are Unsuccessful Avian Invaders Rarer in Their Native Range Than Successful Invaders?**   125
Thomas Brooks

Chapter 8. **A Geographical Perspective on the Biotic Homogenization Process: Implications from the Macroecology of North American Birds**   157
Brian A. Maurer, Eric T. Linder, and David Gammon

Chapter 9. **Global Warming, Temperature Homogenization and Species Extinction**   179
J.L. Green, J. Harte, and A. Ostling

Chapter 10. **The History and Ecological Basis of Extinction and Speciation in Birds**   201
Peter M. Bennett, Ian P.F. Owens, and Jonathan E.M. Baillie

Chapter 11. **Downsizing Nature: Anthropogenic Dwarfing of Species and Ecosystems**   223
Mark V. Lomolino, Rob Channell, David R. Perault, and Gregory A. Smith

Chapter 12. **Spatial Homogenization of the Aquatic Fauna of Tennessee: Extinction and Invasion Following Land Use Change and Habitat Alteration**   245
Jeffrey R. Duncan and Julie L. Lockwood

Chapter 13. **Homogenization of California's Fish Fauna Through Abiotic Change**   259
Michael P. Marchetti, Theo Light, Joaquin Feliciano,
Trip Armstrong, Zeb Hogan, Joshua Viers, and Peter B. Moyle

Contributors   279

Index   283

Chapter 1

Biotic Homogenization: A Sequential and Selective Process

Michael L. McKinney[1] and Julie L. Lockwood[2]

[1]Department of Geological Sciences, and Department of Ecology and Evolutionary Biology
University of Tennessee, Knoxville, TN 37996, USA
[2]Department of Environmental Studies, Natural Sciences II, University of California, Santa Cruz, CA 95064, USA

1. INTRODUCTION

Biotic homogenization is the replacement of local biotas by nonindigenous and locally expanding species that can co-exist with humans (McKinney and Lockwood 1999). Because homogenization often replaces unique endemic species with already widespread species, it reduces spatial diversity at regional and global scales. Biotic homogenization is rapidly increasing to the point that the next geological epoch is sometimes called the "Homogecene" (Guerrant 1992). As noted by Brown (1989),

"Geographically restricted native species with sensitive requirements will continue to have high extinction rates while those widespread broadly tolerant forms that can live with humans, and benefit from their activities, will spread and become increasingly dominant".

Understanding the dynamics of homogenization will be challenging because it subsumes many aspects of the biodiversity crisis including both extinction and species introductions (Hobbs and Mooney 1998). The fossil record shows that biotic homogenization has occurred in the geological past (Roy this volume), and that, in general, two basic forces increase biotic homogenization at any scale: 1) changes in the physical environment ("disturbances") that reduce habitat heterogeneity, and 2) increases in the physical proximity of habitats. The largest mass extinction (thus far), for example, was produced by the unification of earth's continents at the end of the Palaeozoic Era. This not only eliminated over 90% of all species but also created a global biota dominated by a relatively few widespread, broadly adapted "generalist" species (Erwin 1998).

If current trends continue, humans will homogenize the earth's biota to an extent unseen by any previous natural episodes, even the end-Palaeozoic mass extinction. This is because human activities promote habitat homogeneity and habitat proximity to magnitudes far greater than occurred in the geological past (McKinney and Lockwood 1999). Accelerating conversion of wilderness to secondary, farm and urban environments replaces a variety of native habitats with the same already common artificial ones. In addition, increasing global transportation will accelerate the importation of formerly isolated biota to these homogenized environments.

In this chapter, we focus on the dynamics of homogenization. We begin by discussing spatial homogenization, i.e., the replacement of many unique endemic species by a few widespread species. We briefly discuss species-area patterns produced by this process, and then try to identify patterns of extinction and replacement in various nations and states in the U.S. We conclude with evidence that extinction and introduction of species is usually a very non-random process in ecological and taxonomic terms. This is unfortunate for biodiversity conservation because ecological and taxonomic selectivity can enhance the impact of spatial homogenization (McKinney and Lockwood 1999).

2. SPATIAL HOMOGENIZATION

Some insight into the dynamics of homogenization is gained by observing the patterns it produces. A main feature of the current homogenization episode is that humans tend to increase local community (alpha) diversity by the introduction of many non-native species, but simultaneously decrease regional and global diversity because the same species are introduced in many places (Harrison 1993, Noss and Cooperrider 1994, Brown 1995, Marchetti this volume, Duncan and Lockwood this volume). Habitat homogeneity decreases the slope of species-area curves, as when continents have lower slopes than islands (Rosenzweig 1995). It is therefore not surprising that human activities such as agriculture, which replace a variety of natural habitats with fewer, more uniform habitats, produce lower slopes in species-area curves than existed before human impacts (Flather 1996; Thiollay 1995; McKinney 1998).

In its simplest expression, spatial homogenization can be viewed as a process of extinction via habitat loss followed by biotic range expansion occurring via habitat gain (often through the introduction of non-native species). "Losers" are those species which occupy habitats frequently destroyed by humans (e.g., old-growth forests, wetlands) and "winners" are those that, through past evolutionary adaptations, occupy habitats that humans often create (e.g., secondary forest, urban areas). Spatial homogenization therefore increases as the proportion of "winners" shrinks and that of "losers" increases.

A preliminary compilation shows that the potential for extensive biotic homogenization is very high because only 5-29% of regional species benefit from various human activities versus a much larger percentage (usually well over 50%) of species decline from human activities (McKinney and Lockwood 1999). These percentages indicate that global homogenization (the "Homogecene") has barely begun: only 1-2% of most birds and mammals (the best-studied groups) are considered to be successfully introduced somewhere in the world (Long 1981). If regional trends are any gauge (e.g., Hobbs and Mooney 1998), the number of successfully introduced birds will almost certainly increase. Assuming that beneficiary species continue to be drawn from a pool of naturally widespread and abundant species (Goodwin et al 1999; Brooks this volume), spatial homogenization is further enhanced.

2.1   Spatial Homogenization as a Sequential Process

The spatial homogenizing process can potentially follow many pathways. For example, extinction may open niches to permit successful invasion, invasion may cause extinction of native species via biotic interactions, or habitat change may both cause extinction and allow invasion. The literature on species introductions is largely anecdotal but it shows that all three of these pathways, as well as others such as introductions without extinctions, have often occurred (Williamson 1996). Hopefully, future work will quantitatively document which pathway (or sequence of events) is most common in the extinction and replacement process, and under what circumstances a given sequence occurs.

Among the evidence already assembled, a few studies stand out. For islands, Case (1996) found a significant correlation between the number of extinct birds per island and the number of birds introduced to that island. He suggested that this correlation occurred because habitat loss caused extinction of native species and also promoted establishment of non-native species.

Do continental ecosystems also show a correlation between number of extinct species and number of introduced species? We might suspect so because, as Case (1996) noted, both extinctions and introductions tend to increase with increasing human presence. However, the correlation might be weaker because introduced species have historically played a lesser role as a cause of extinction and threat on continents than islands. For example, alien species threaten over 98% of imperilled bird and plant species on Hawaii compared to less than 50% of imperilled birds and plant species in the continental U.S. (Wilcove et al. 1998). This could potentially reduce a direct causal linkage between extinction and introductions in continental ecosystems.

To test this, we regressed the number of successfully introduced plant species versus the number of threatened plant species in a continental region. For continental regions, we used available data for nations of the world and states in the U.S. Threatened plant species per nation were taken from Walter and Gillett (1998) and for each state from the Nature Conservancy Natural Heritage network website (www.tnc.org). Numbers of introduced plants for each nation are from Vitousek et al. (1996) and for each state from Mac et al. (1999).

The results (Fig. 1) indicate that, as with islands, continental ecosystems show a significant correlation between number of species threatened and number of species successfully introduced. The regression for these data (n = 25) is highly significant (F = 125; p < 0.00001). The regression equation, y = 0.71x + 181, explains 84% of the variation in the variables (R^2 = 0.84). Interestingly, the slope (95% confidence interval for slope ranges from 0.58 to 0.84) is almost exactly the same as that found by Case (1996) for a similar regression of island birds. This would indicate that, on average, species committed to extinction have tended to outpace introductions.

Because of the strong regression effect of one point (the U.S.), we redid the regression without that point. The regression for those data (n = 24) is still highly significant (F = 30.0; p < 0.0001). The regression equation, y = 0.49 + 278, explains about 54% of the variation. The slope declines substantially when the U.S. is excluded (95% confidence interval of 0.30 to 0.67), indicating a greater average increase in threatened species over introductions. Even with the removal of other outliers, South Africa and Mexico (Fig. 1), the regression slope continues to be less than one (0.79, 95% confidence interval = 0.62 to 0.96) indicating that threatened species tend to outnumber successfully introduced species in most areas.

In discussing his results, Case (1996) suggests that this correlation between extinction and introduction is related to habitat loss of native species followed by habitat gain of introduced species as human colonization intensifies. On islands, for example, human occupation produces rapid loss of native habitat, causing extinctions. Urban, farm and other human-dominated habitats that replace native habitat typically have the highest proportions of non-native species of any biota (Kowarik 1995; Blair this volume). In Case's (1996) scenario, highly disturbed islands will therefore have high numbers of extinctions and introductions while less disturbed islands will tend to have low numbers of both.

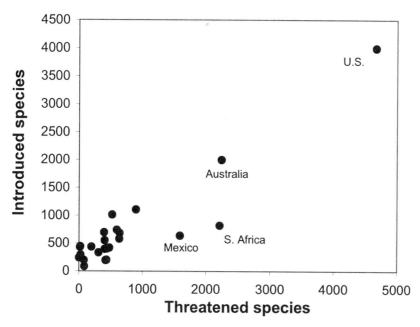

Figure 1. Correlation between number of threatened plant species and number of established non-native plant species for various regions (states and nations) of the world.

A testable implication of this sequential process of homogenization is that areas with historically low human population densities and thus short histories of human impacts should have many threatened species and relatively few introduced species. In contrast, highly populated areas such as European nations with a long history of extensive human impacts, should have many more introduced species for at least two reasons: 1) increasing human population density increases the number of species introduced (Chown et al. 1998; Lonsdale 1999), 2) the widespread occurrence of human-dominated ("disturbed") habitats that are hospitable to non-native species. Conversely, nations with a long history of human impacts may have relatively low numbers of threatened species because many species that are prone to extinction from human activities will have already disappeared. This is the "extinction filter" of Balmford (1996). More quantitatively, these patterns predict that, for a given area, the ratio I/T (where I = number of species introduced and T = number of species threatened) will increase with increasing human population density.

To test this, we used the plant threat and introduction data for each nation and state above (Fig. 1) to create a single ratio variable

$$D = \frac{(I-T)}{O} \qquad \text{(EQUATION 2.1.1)}$$

where O = original number of native species in the nation or state. (O was estimated as current species diversity minus number of introduced species.) The ratio variable, D, is thus the net gain or loss of species after introduction of non-native species and extinction of threatened native species, expressed as a percentage of the original species diversity. For each area (nation or state), D was plotted against the human population density of that area (e.g., total population of a nation divided by the area of that nation). If human impacts are indeed sequential in the way described, then nations and states with low human population densities would tend to have relatively low or even negative D as more species are threatened than are introduced. In contrast, increasing population densities would increase D.

The results support the predicted net gain of species with increasing human population density (Fig. 2). Considering the many potential biases and other problems in estimating numbers of threatened and introduced species in different states and nations, the relationship is perhaps striking in how well it follows the predicted pattern. The regression ($y = 0.0011x - 0.054$) is highly significant ($F = 23.6$; $p < 0.00015$), and explains 56% of the variation. The positive slope (95% confidence interval = 0.00062 to 0.016) indicates increasing diversity gain with population density. Also, areas with low population densities (e.g., Mexico, U.S., Tanzania, S. Africa, Australia) tend to fall below zero, indicative of a depauperate biota (threatened species > introduced species). In contrast, areas with high population densities, mainly European nations, tend to have the largest diversity gain (introduced species >> threatened native species). This pattern (Fig. 2) quantifies the sequential enrichment of local biota by increasing species introductions and the simultaneous loss of native species, as human population density increases.

Many studies have documented a strong positive correlation between the number of non-native species and human population size (Chown et al. 1998; Lonsdale 1999). This would account for part of the patterns found here because

$$D = \frac{(I-T)}{O} = \frac{I}{O} - \frac{T}{O} \qquad \text{(EQUATION 2.1.2)}$$

so that I/O is documented to increase with increased population density. But much less has been documented about T/O, the second factor determining D. Does T/O actually decrease with increasing population

1. Biotic Homogenization: A Sequential and Selective Process

density, as predicted by Balmford's (1996) "extinction filter"? A plot of T/O versus population density (Fig.3) for the data used earlier does support this idea. The relationship is apparently non-linear, so the y-axis (% threatened) has been logarithmically transformed, indicating a rapid loss of extinction-prone species with initial increase in human population density. The regression parameters are: $\ln y = -0.016x - 1.7$ ($R^2 = 0.44$; $F = 14.9$, $p < 0.0013$). The relationship is inverse (95% confidence interval of slope = -0.007 to -0.03).

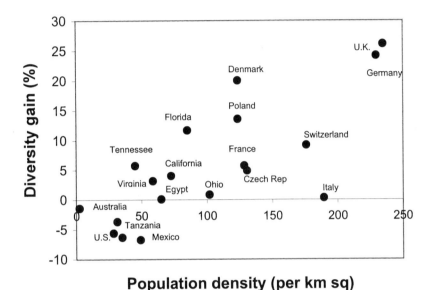

Figure 2. Plot of plant diversity gain of a region versus the human population density of the region. Significant positive relationship indicates increasing gain in local diversity with human population density.

This inverse exponential relationship between threatened species and population density indicates that the percentage of threatened species decreases by roughly half for each increase of 44 persons per square km (ln 2/1.6% = 44). The general exponential pattern of decline is consistent with many studies of extinction lag times from habitat fragmentation that show an initially high rate of loss followed by progressively slower rates (Brooks et al. 1999).

Our results contrast with Thompson and Jones (1999) who found that increasing human population density causes an increase in local plant extirpations in Britain. This discrepancy apparently results from the very coarse scale of our data compared to the very fine scale of the British study.

Our coarse comparison identifies very broad differences in the impact of humans between areas with very distinct histories. Eventually, one would expect that the accumulation of local extirpations in Britain would produce local floras that also show the "extinction filter" effect of Balmford (1996).

2.2 Local Spatial Homogenization: Urban-rural Gradient Turnover

Given the role of scale, it is important to analyze spatial patterns of local extinction and replacement along gradients that range from low-diversity urban habitats to high-diversity natural habitats (Blair this volume; Beissinger and Osborne 1982; Limburg and Schmidt 1990). The predictive utility of this approach is that it shows what will happen to a biota as urban habitats expand and replace natural and slightly modified habitats.

To measure this human impact on net diversity, we applied equation 2.1.1 on the urban gradient bird data of Blair (this volume). These data describe the spatial appearance and disappearance of birds across a gradient ranging from relatively undisturbed habitat (preserves) to a highly disturbed habitat (business district) in two different parts of the U.S (Ohio and California). The formula was modified to measure the net gain or loss in diversity (D) as one moves from the preserve across the spatial gradient to increasingly disturbed habitats. In the formula, I becomes the number of species that appear in a habitat (i.e., did not occur in the preserve), T is the number of species that disappear from a habitat (i.e., did occur in the preserve) and O is the original number of species in the preserve. For example, of 21 bird species in the preserve near Palo Alto, California, seven of these disappear in the adjacent open space while eight new birds appear in the open space (Blair this volume). This gives a net gain of 5% or 0.05 (= (8 − 7)/21). The next-most disturbed area along the gradient (golf course) loses 10 preserve birds but gains 17 new ones for a net gain of 33%.

The general pattern observed across the disturbance gradients of both the California and Ohio data reveals an "intermediate-disturbance" diversity peak (Fig. 4). Specifically, golf courses tend to have the most species, about 1/3 more than in the preserve. In contrast, highly urbanized areas (business districts, research parks) have the least diversity with less than 70% of the species number of preserves. This same "intermediate-peak" pattern is also found in similar gradients with butterflies (Blair this volume) and plants (Kowarik 1995). Importantly, most of the diversity gain consists of species living in nearby habitats that move into the moderately disturbed areas. Only the most highly disturbed habitats contain introduced species from other continents in these data. However, homogenization still occurs at

1. Biotic Homogenization: A Sequential and Selective Process

intermediate disturbance levels because of the expansion of "disturbance-adapted" local species replacing other local species that cannot adapt to disturbance.

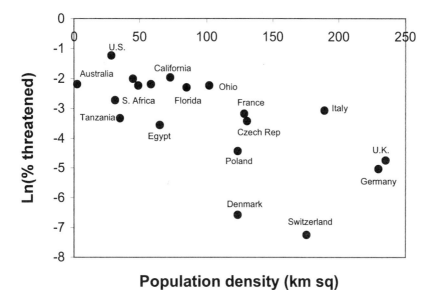

Figure 3. Plot of ln (% threatened plant species) versus human population density for that region. Significant negative correlation indicates a non-linear decrease in threatened flora with increasing time of human occupation.

The processes underlying these urban gradient patterns has been discussed elsewhere (Blair this volume, Kowarik 1995). Primary factors include: the replacement of vegetation by concrete in highly disturbed areas to produce a net loss of species diversity, and the increasing variety of plants, such as ornamentals, in moderately disturbed areas that increase the habitat diversity beyond that found in undisturbed areas. Also, Case (1996) notes that moderately disturbed areas usually consist of a mosaic of undisturbed and disturbed habitats and that species-area effects show that a higher diversity of species can coexist in a given area in such conditions. The result is that human activities tend to increase local community diversity (the y-intercept on the species-area curve) in most areas (because they are only moderately disturbed) while decreasing overall regional diversity (the slope on the species-area curve; Harrison 1993; Brown 1995; McKinney 1998).

Interestingly, these spatial gradient patterns are matched by similar temporal patterns. A number of studies have found that, following intensive disturbance such as urbanization, the number of bird species inhabiting an area will increase through time (Vale and Vale 1976; Aldrich and Coffin 1980). This is because the growth of vegetation, especially ornamental plants, creates new habitat after native habitat has been destroyed. If the highly disturbed area is left alone to regenerate, it may (through time) pass through the same diversity changes observed in the spatial gradient (Fig. 4) from highly disturbed to relatively undisturbed, with maximum diversity in an intermediate level of disturbance.

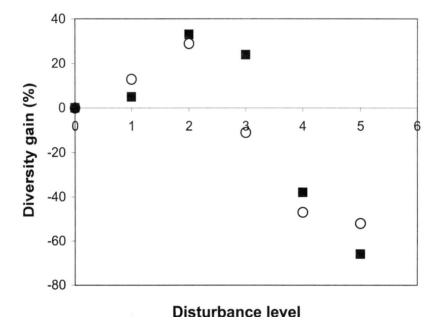

Figure 4. Plot showing increase in local bird diversity with intermediate levels of disturbance. Data from Blair (this volume). Squares = California data; diamonds = Ohio data. Disturbance is ordinal scale: 1 = preserve, 2= open space, 3 = golf course, 4 = residential, 5 = apartments/research park, 6 = business district.

3. TAXONOMIC AND ECOLOGICAL SELECTIVITY ENHANCES HOMOGENIZATION

Spatial homogenization can be enhanced if the winning and losing species are not randomly distributed among taxonomic and ecological categories (McKinney 1997, Lockwood et al. 2000).

1. Biotic Homogenization: A Sequential and Selective Process

All species are not equally likely to become extinct. Extinction-biasing traits such as large body size, low fecundity, or rarity are often clumped within higher taxa such as families or genera (McKinney and Lockwood 1999). This clumping of traits (i.e. selectivity) is apparently a very common result of the non-random way that traits are shared among closely related species (McKinney 1997). Clumping of extinction-biasing traits means that the unique morphological or behavioral diversity that these higher taxa represent will be lost at a much higher rate than if extinction happened at random. The rate of loss can be especially severe if extinctions are concentrated within species–poor taxa. Extinction selectivity operates at the global and local scale. Table 1 lists groups that are extinction-prone (losers) as well as some of the biasing traits reported. As illustrated by the parrots, the same group may be extinction-prone at global and local scales, and often for similar reasons. Globally, the parrot family (Psittacidae) has many more threatened species than expected when compared to other bird families (Bennett and Owens 1997). Bennett and Owens (1997) attribute this selectivity to parrot's large body size and low fecundity. These traits make parrots susceptible to over-exploitation and habitat loss. Locally, parrots in Brazil are more likely to decline after deforestation as compared to other Brazilian forest birds (Goerck 1997). Goerck (1997) suggests that the larger body size of parrots and their tendency toward frugivory make them vulnerable.

Species are not equally likely to become successfully established outside of their native range either. These 'winning' species include those that expand locally after a disturbance and those that find themselves on different continents as exotics. As with extinction-biasing traits, such clustering of winning species in certain families can be attributed to the non-random way that traits are shared among closely related species (McKinney 1997). Interestingly, the list of traits that bias species toward extinction are the opposite of those purported to bias species toward expanding their range (Lockwood 1999, McKinney and Lockwood 1999). This can be seen in Table 1. Groups that have expanded their range either locally or globally are listed, as are some of the biasing traits reported. The winners in Table 1 include groups with generalist adaptations (e.g., non-plant dependent webs) or high fecundity rates (e.g., weedy grasses).

It is not always the case that groups can be clearly placed into winners or losers. Among birds, several families are simultaneously considered extinction-prone and invasion-prone (Lockwood et al. 2000). These include the ducks (Anatidae), pheasants (Phasianidae), parrots, and pigeons and doves (Columbidae). Often the same traits that make a group susceptible to extinction make it favored for introduction. Pheasants, for example, are typically large-bodied and have ornate plumage. These traits seem to bias this species toward over-explotation from hunting. However, these traits also bias this group towards intentional transport and release outside of their native range – usually for hunting purposes. As far more Pheasants are threatened than successfully introduced, the end result is fewer pheasant species that are re-distributed over the globe (Lockwood et al. 2000).

Table 1. Ecological and taxonomic patterns among winners and losers (modified from McKinney and Lockwood 1999)

Over-represented groups	Biasing traits	Citations
Losers		
Parrots	Large size, low fecundity	Bennett and Owens 1997
Apes	Large size, low fecundity	Russell et al. 1998
Sumatran Babblers	Large size, forest specialist	Thiollay 1995
Chihuhuan desert Cacti	Small range	Hernandez and Barcenas 1996
Tropical reef fish	Large maximum size	Jennings et al. 1999
Brazilian Parrots	Large size, frugivory	Goerck 1997
Winners		
Grass as world-wide weed	Rapid dispersal and growth	Daehler 1998
Legumes as natural area invaders world-wide	Broadly tolerant	Daehler 1998
Appalachian stream fish	Pit-spawning, silt tolerant	Jones et al. 1999
Amazonian Hylid frogs	Breed in temporary ponds	Pearman 1997
South African Wolf spiders	Plant independent webs	Ingham and Samways 1996
Ducks world-wide	Transported as game	Lockwood 1999

This points to a critical observation concerning the prevalence of taxonomic selectivity, especially as it is related to conservation planning. Taxonomic clumping of traits can occur at any point along the route towards extinction or invasion. For example, avian higher taxa that contain many successfully introduced species most often have traits that promote intentional transport rather than establishment (Lockwood 1999). Similarly, Daehler (1998) found that plant higher taxa that invade natural areas posses a different set of traits than taxa that are considered agricultural weeds. Thus, when trying to predict which species will become invaders it is important to distinguish between traits that pre-dispose groups to 1) transport, 2) establishment, and 3) spread. The same is likely true for the steps that lead to extinction. Biasing traits may, or may not, be similar among steps. Understanding the differences can inform management decisions by concentrating limited resources not only toward one group but also toward a particular stage in the process.

A twist on the phenomena of taxonomic homogenization is hybridization (Daehler this volume). Hybridization between sister taxa can produce a homogenous gene pool, which occasionally leads to the functional loss of one of the two hybridizing species (Rhymer and Simberloff 1996). These sister taxa have not evolved reproductive isolation, however, this is often unknown because in natural situations a barrier (e.g., expanses of uninhabitable forest or an ocean) separates them. There are a surprising number of cases where this occurs including the hybridization between native and exotic *Spartina* species in the San Francisco bay (Daehler and Strong 1997) and between Golden-winged by the Blue-winged warblers in the southeastern U.S. (Gill 1997). It is not necessary that hybridization occur between closely related species (Gill 1998). For example hybridization across genera is a common phenomena among birds (called Parkes' Paradox; Gill 1998, Parkes 1978). Thus, hybridization could increase homogenization by decreasing richness of one genera in favor of the expansion of another.

This form of taxonomic homogenization increases as spatial homogenization increases. The incidence of hybridization is growing as more sister taxa are artificially brought into secondary contact (Gill 1980). Spatial homogenization thus increases secondary contact, and hybridization, from 1) changing landscape features such that habitat configurations no longer provide a natural barrier between sister taxa (e.g., conversion of forests into rangeland) and 2) increased human transportation and introduction of species into the ranges of reproductively compatible natives (e.g., exotic fishes interbreeding with natives).

Homogenization can also be enhanced by selectivity among ecological categories. Specifically, spatial homogenization in the future biosphere (as in past prolonged mass extinctions) could be exaggerated by the local replacement of ecological specialists with the same widespread generalists and opportunists. This produces ecological homogenization by the replacement of many (often more complex) functional and ecological systems by the same few simpler ones. Brooks (this volume) explores ecological selectivity among successful versus failed introduced birds. Brooks suggests that rarity (as indexed by being an ecological specialist, having a small range size, or being scarce) should decrease a species' chance of successful introduction into a novel environment. Brooks finds that being an ecological specialist is the only form of rarity that translates into a low probability of successful introduction. Corresponding evidence that ecological specialists are more likely to go extinct, has been shown for many groups (McKinney 1997). Lomolino (this volume) explores another expression of ecological selectivity, namely that small bodied vertebrates are more likely to be winners than their large bodied cousins.

It is not yet clear if we should necessarily see a corresponding phylogenetic or taxonomic pattern resulting from ecological selectivity. Webb et al. (this volume) indicate that while extinction risk is correlated to rarity, and there is taxonomic selectivity among species at risk for extinction, there is no phylogenetic pattern for rarity. There are at least two reasons why ecological and taxonomic selectivity may not be tightly linked. First, ecological traits that are selected for or against may not be heritable to the extent that they are conserved within lineages. Second, ecological changes that cause extinctions (or local extirpations) are often profound (Duncan and Lockwood this volume, Marchetti this volume). The process of spatial homogenization often involves the replacement of entire ecosystems. Regionally unique ecosystems are removed (e.g., freshwater springs, wetlands) and in their place a 'new' more common ecosystem is created (e.g., farmland, suburban housing). It is within these new ecosystems that exotic species establish or local species expand their ranges. This turnover may, or may not, translate into a taxonomic or phylogenetic pattern depending on 1) whether historic speciation processes have produced adaptations to both types of environments within the same family and 2) what propagules find their way to the 'new' ecosystem.

4. SUMMARY

Biotic homogenization can be seen as the principle by-product of a growing human population. However, it is not the presence of humans in the landscape that promotes homogenization, it is the cultural and economic impacts of humans. For example, Thompson and Jones (1999) show that British vice-counties with low human populations have high proportions of their land devoted to agriculture. These vice-counties are not the principle threat to native plant species richness in Britain whereas increasingly urbanized vice-counties are. Thus, there is something specific to urbanization that threatens native plants. Similarly, in urban-rural gradients, it is the rural areas (not the preserves) that have the highest species richness; whereas the urban areas host the fewest, and the largest number of non-native, species. This points to an over-arching solution to projected increases in biotic homogenization; a more ecologically thoughtful approach to urbanization and the conservation of urban open-space (Theobald et al. 1997, Collinge 1996, Stalter et al. 1996).

There are other homogenizing impacts of humans that are more difficult to negotiate. One of the more obvious impacts is the increased ability of species to overcome historically isolating boundaries. This reflects two aspects of human culture. First, humans tend to purposefully and inadvertently transport species around the world (Williamson 1996). Second, human impacts are increasing to the point that formerly separated floras and faunas are coming into artificial contact. These two factors tend to increase homogenization through hybridization, increased 'mixing' of biotas, and because the species that win (i.e., species being transported or prospering after land-use changes) come from non-random taxonomic and ecological groups. Efforts to curb these effects are complicated by international politics, landholder rights, and the complexity of the underlying ecological patterns. Some of the apparent intractability surrounding the solutions to human cultural impacts is an unwillingness (or simple neglect) of ecologists to consider how cultural biases translate into evolutionary and ecological trends.

Finally, it is increasingly apparent that addressing conservation problems requires an understanding of how environmental change effects both the winners and the losers. In some circumstances the two groups are directly linked, such as when an exotic predator feeds on native species. In others the effects are more subtle and likely result from a common underlying mechanism such as changes in land use that replicate particular habitats across the landscape (e.g., lawns). Research that explicitly ranks the possible mechanisms of homogenization is warranted, as is the recognition of varied human impacts. It is our hope that this volume provides a coherent beginning to research on biotic homogenization.

REFERENCES

Aldrich, J.W. and R.W. Coffin. 1980. Breeding bird populations from forest to suburbia after thirty-seven years. American Birds 34(1):3-7.

Balmford, A. 1996 Extinction filters and current resilience: the significance of past selection pressures for conservation biology. Trends in Ecology and Evolution 11:193-197.

Beissinger, S.R. and D.R. Osborne. 1982. Effects of urbanization on avian community organization. Condor 84:75-83.

Bennett, P.M. and I. Owens. 1997. Variation in extinction risk among birds: chance or evolutionary predisposition? Proc. R. Soc. London Ser.. B. 264:401-408

Brooks, T.M., S.L. Pimm and J.O. Oyugi. 1999. Time lag between deforestation and bird extinction in tropical forest fragments. Conservation Biology 13:1140-1150.

Brown, J.H. 1989. Patterns, modes and extents of invasions of vertebrates. In: Biological Invasions: a Global Perspective (Drake, J. et al., eds), John Wiley & Sons, pp. 85-109.

Brown, J.H. 1995. Macroecology, University of Chicago Press

Case, T.J. 1996. Global patterns in the establishment and distribution of exotic birds. Biological Conservation 78:69-96.

Chown, S.L., N.J. Gremmen, and K.J. Gaston. 1998. Ecological biogeography of Southern Ocean islands: Species-area relationships, human impacts and conservation. American Naturalist 152:562-575.

Collinge, S.K. 1996). Ecological consequences of habitat fragmentation: implications for landscape architecture and planning. Landscape and Urban Planning 36:59-77.

Daehler, C.C. 1998 The taxonomic distribution of invasive angiosperm plants: ecological insights and comparison to agricultural weeds, Biol. Conserv. 84:167-180

Daehler, C.C. and D.R. Strong. 1996. Status, prediction and prevention of introduced cordgrasses (Spartina spp.) in Pacific estuaries, USA. Biological Conservation 78:51-58.

Erwin, D.H. 1998. The end and beginning: recoveries from mass extinctions, Trends Ecol. Evol. 13:344-349

Flather, C.H. 1996. Fitting species-accumulation functions and assessing regional land use impacts on avian diversity. Journal of Biogeography 23:155-168

Gill, F. 1980. Historical aspects of hybridization between Blue-winged and Golden-winged Warblers. Auk 97:1-18.

Gill, F. 1997. Local cytonuclear extinction of the Golden-winged Warbler. Evolution 51:519-525.

Gill, F. 1998. Hybridization in birds. Auk 115(2):281-283.

Goerck, J.M. 1997. Patterns in rarity in the birds of the Atlantic forest of Brazil. Conserv. Biol. 11:112-118

Goodwin, B.J., A. J. McAllister, and L. Fahrig. 1999. Predicting invasiveness of plant species based on biological information. Conservation Biology 13: 422-426.

Guerrant, E.O. 1992. Genetic and demographic considerations in the sampling and reintroduction of rare plants. In: Conservation Biology (P.L. Fiedler and S. Jain, eds.) Chapman and Hall, London, pp. 321-344.

Harrison, S. 1993. Species diversity, spatial scale, and global change. In: Biotic Interactions and Global Change (Kareiva, P., Kingsolver, J. and Huey, R., eds), Sinauer, pp. 388-401.

Hernandez, H.M. and R. Barcenas 1996. Endangered cacti in the Chihuahuan Desert, Conserv. Biol. 10:1200-1209

Hobbs, R.J. and H. Mooney. 1998. Broadening the extinction debate: population deletions and additions in California and Western Australia, Conserv. Biol. 12:271-283

Ingham, D. and M. Samways. 1996. Application of fragmentation and variegation models to epigaeic invertebrates in South Africa, Conserv. Biol. 10:1353-1358

Jennings, S., J.D. Reynolds, and N.V.C. Polunin. 1999. Predicting the vulnerability of tropical reef fishes to exploitation with phylogenies and life histories. Conservation Biology 13(6):1466-1486.

Jones, E.B.D. III, G.S. Helfmen, J.O. Harper, and P.V. Bolstad. 1999. Effects of riparian forest removal on fish assemblage4s in southern Appalachian streams. Conservation Biology 13(6):1454-1465.

Kowarik, I. 1995. On the role of alien species in urban flora and vegetation. In: Plant Invasions: General Aspects and Special Problems, P. Pysek, K. Prach, M. Rejmanek, and M. Wade, eds., Academic Pub., Amsterdam. Pp: 85-103.

Limburg, K.E. and R.E. Schmidt. 1990. Patterns of fish spawning in Hudson River tributaries: response to an urban gradient? Ecology 71(4):1231-1245.

Lonsdale, W.M. 1999. Global patterns of plant invasions and the concept of invasibility. Ecology 80:1522-1536.

Lockwood, J.L. 1999. Using taxonomy to predict success among introduced avifauna: relative importance of transport and establishment. Conservation Biology 13(3):560-567.

Lockwood, J.L., T.M. Brooks, and M.L. McKinney 2000. Taxonomic homogenization of the global avifauna. Animal Conservation 3:27-35.

Long, J. 1981. Introduced birds of the world. David & Charles, London.

Mac, M.J., P.A. Opler, C.P. Haecker, and P. D. Doran. 1998. Status and Trends of the Nation's Biological Resources. U.S. Dept Interior, Reston, Virginia.

McKinney, M.L. 1997. Extinction vulnerability and selectivity: combining ecological and paleontological views, Annu. Rev. Ecol. Syst. 28:495-516

McKinney, M.L. 1998. On predicting biotic homogenization: species-area patterns in marine biota, Glob. Ecol. Biogeogr. Lett. 7: 297-301

McKinney, M.L. and J.L. Lockwood. 1999. Biotic homogenization: a few winners replacing many losers in the next mass extinction. Trends in Ecology and

Evolution 14: 450-453.

Noss, R. and A.Y. Cooperrider. 1994. Saving Nature's Legacy. Island Press, Washington DC.

Parkes, K.C. 1978. Still another parulid intergeneric hybrid (Mniotilla X Dendroica) and its taxonomic and evolutionary implications Auk 95:682-690.

Pearman, P.B. 1997. Correlates of amphibian diversity in an altered landscape of Amazonian Ecuador, Conserv. Biol. 11:1211-1225

Rhymer, J.M. and D. Simberloff 1996. Extinction by hybridization and introgression. Annual Review of Ecology and Systematics 27:83-109.

Rosenzweig, M.L. 1995. Species Diversity in Time and Space. Cambridge University Press, Cambridge.

Russell, G.J., T.M. Brooks, and M. McKinney. 1998. Decreased taxonomic selectivity in the future extinction crises. Conservation Biology 12:1365-1376.

Stalter, R., M.D. Byer, J.T. Tanacredi. 1996. Rare and endangered plants at Gateway National Recreation Area: a case for protection of urban natural areas. Landscape and Urban Planning 35:41-51.

Theobald, D.M., J.R. Miller, and N.T. Hobbs. 1997. Estimating the cumulative effects of development on wildlife habitat. Landscape and Urban Planning 39:25-36.

Thiollay, J-M. 1995. The role of traditional agroforests in the conservation of rain forest bird diversity in Sumatra, Conserv. Biol. 9:335-353

Thompson, K. and A. Jones. 1999. Human population density and prediction of local extinction in Britain. Conservation Biology 13(1):185-189.

Vale, T. R. and G.R. Vale. 1976. Suburban bird populations in west-central California. Journal of Biogeography 3:157-165.

Vitousek, P.M. et al. 1996. Biological invasions as global environmental change, Am. Sci. 84:468-478

Walter, K.S. and H.J. Gillett. 1998. 1997 IUCN Red List of Threatened Plants. World Conservation Union, Gland, Switzerland.

Wilcove, D.S., D. Rothstein, J. Dubow, A. Philips, and E. Losos. 1998. Quantifying threats to imperiled species in the United States. BioScience 48:607-615.

Williamson, M. 1996. Biological invasions, Chapman & Hall .

Chapter 2

Biotic Homogenization: Lessons from the Past

Kaustuv Roy[1] and Jeffrey S. Kauffman[2]
[1]*Department of Biology, University of California, San Diego,La Jolla, CA 92093-0116, USA*
[2]*Department of Biology, Indiana University, Bloomington, IN 47405-3700, USA*

1. INTRODUCTION

The ongoing process of biotic homogenization has two distinct components, extinction of species and invasion of species into new areas (Brown and Lomolino 1998; McKinney 1998). Both invasions and extinctions are an integral part of the evolutionary process (Williamson 1996), and the fossil record provides a rich source of data on the effects of each of these processes on species assemblages at a variety of spatial and temporal scales. These data allow a glimpse into how the processes of invasions and/or extinctions unfold over timescales beyond the reach of ecological data, and provide a historical framework for looking at ongoing patterns of biotic homogenization. The relevance of fossil data to the study of ongoing species extinctions have been extensively discussed (Jablonski 1995; McKinney 1997), but invasions have arguably received less attention. Here we will focus on patterns of invasions, past and present.

Paleobiological data show that the geographic ranges of species can change rapidly in response to environmental fluctuations and that such changes occur at many different temporal scales, from decades to centuries to millions of years (Valentine and Jablonski 1993; Roy et al. 1996). Thus the patterns of invasions preserved in the fossil record cover a variety of temporal and spatial scales and examples range from the responses of taxa to Pleistocene climatic fluctuations, the invasion of the North Atlantic ocean by North Pacific mollusks in the Pliocene (Vermeij 1989 1991; Reid 1990), the Great American Interchange following the formation of the Isthmus of Panama (Marshall et al. 1982), to entire continental biotas coming together with the formation of Pangea during the Paleozoic (Sheehan 1988).Thus biological invasions are an integral part of the long term dynamics of ecosystems, and the examples preserved in the fossil record should help us better understand the process as well as the consequences of biological invasions (Elton 1958; di Castri 1989; Valentine and Jablonski 1993; Lodge 1993). Despite the differences in temporal and spatial scales, data from the fossil record have already shown that some general characteristics are shared by many invasions, past and present. One example is the asymmetry of invasion patterns, with low diversity communities being more invasion prone. Such a trend is well documented for modern invasions (Williamson 1996) and for biotic interchanges during the Cenozoic (Vermeij 1991; Valentine and Jablonski 1993). Similarly, the observation that extinctions make biotas more susceptible to invasion is also supported by data from modern case studies (Case 1996) and the fossil record (Vermeij 1991). While such similarities are very informative we also need to ask whether there are aspects of human-mediated invasions that separate them from the natural ones in the past. That's the question we address here.

In addition to those cited above, proposed generalizations about the invasion process include the observation that only a small percentage of introduced species actually become established in the new habitats, and that invasive species tend to be a non-randomly distributed among higher taxa (Lodge 1993; Williamson 1996; Daehler 1998; Pysek 1998; Lockwood 1999). Generalizations about traits that make species better at invading new areas include high intrinsic rates of natural increase, large geographic range and/or abundance in native habitat, high genetic variability, high dispersal rate, climate and habitat matching, and taxonomic isolation (see Lodge 1993; Daehler and Strong 1993; Williamson 1996). There is considerable debate about the validity of some of these generalizations and most have not yet been subjected to extensive statistical tests (Lodge 1993; Williamson 1996). In addition, it is not clear if they apply across different temporal and spatial scales, and for both natural and human-mediated events. Here we focus on three generalizations that are particularly relevant to the issue of biotic homogenization but have not seen a lot of previous work. By comparing Pleistocene patterns with the present day we argue that while ongoing human induced invasions have a number of similarities with the natural events preserved in the fossil record, they also appear to differ in some important ways. These differences may prove to be important in understanding the process of homogenization. Since comparisons of past and present patterns of invasion cannot be undertaken at the level of individual species or populations, we will focus on emergent statistical patterns that might characterize invasions in general. Such an approach can be informative and has been advocated (Gilpin 1990; Lodge 1993; Brown 1995; Williamson 1996).

2. THREE GENERALIZATIONS

We start by looking at three aspects of invasions that influence the process of homogenization: (1) geographic range of invasive species, (2) taxonomic composition of invasive species, and (3) rates of invasions.

2.1 Geographic range and invasion success

Case studies of ongoing invasions suggest that geographic range of a species in its native habitat is an important predictor of invasion success, with more widespread species being better invaders (Moulton and Pimm 1986; Crawley 1987; Ehrlich 1989; Daehler and Strong 1993). Is large geographic range also a good predictor of invasion success during natural invasions, such as those driven by past climate change? While Pleistocene data are well suited for addressing this question, few such analyses are available in the literature.

One study that did examine the role of geographic range in invasions during the Pleistocene used data for the shallow marine molluscan fauna of California (Roy et al. 1995). The Pleistocene molluscan fossil record of California contains many species that today live only to the north or south of the Californian province, defined as the region between 28°N and 34.5°N (Roy et al. 1995 1996; Fig. 1). Species exhibiting such extensive inter-provincial range shifts are termed extraprovincial species and form about 12% (90 out of 750) of the total diversity. Roy et al. (1995) showed that overall the extraprovincial species of Californian mollusks have significantly larger latitudinal ranges compared to the non-extraprovincial species (i.e. the species pool from which they were drawn). While this is consistent with the pattern seen in present day invasions, Roy et al. (1995) also showed that in the Pleistocene case the differences in ranges held only for the species that extended their range northwards (or southern extraprovincial species) during the climatic fluctuations (Fig. 2). Species that extended their ranges southwards (i.e. northern extraprovincial species) did not have significantly longer latitudinal ranges compared to the rest of the species pool (Fig. 2). Roy et al. (1995) attributed this difference to the fact that for marine species, contacts between major water masses such as those forming the provincial boundaries, tend to be more effective barriers for southern species moving north than for northern species moving south. For marine mollusks, species that are latitudinally wide ranging also tend to be physiologically the most tolerant, and hence such long ranging southern species can successfully cross biogeographic boundaries and preferentially colonize new areas following climate change (Roy et al. 1995). Since biogeographic barriers are leaky filters for southern range end points, geographic range and its correlates matter less for species moving south.

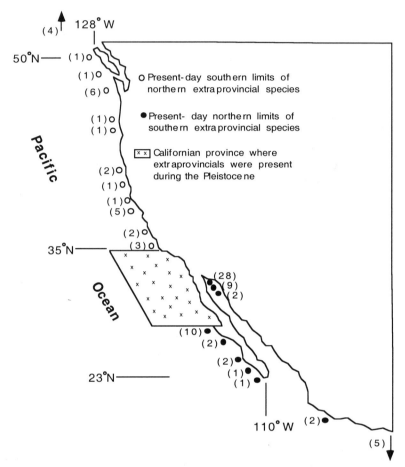

Figure 1. Modern distribution of Pleistocene extraprovincial marine molluscan species along the eastern Pacific margin. Numbers in parentheses are numbers of extraprovincial species whose ranges end at that latitude. Arrows indicate that southern range limits of four northern extraprovincial species and northern limits of five southern extraprovincial species fall outside the area shown. (modified from Roy et al. 1995).

Is the pattern revealed by the marine mollusks a general Pleistocene pattern, one that also applies to terrestrial animals as well as plants? The answer is not clear from the literature since comparable analyses have not been done using terrestrial data. However, a preliminary result for the Pleistocene land snails of eastern North America can be obtained using the range maps in Hubricht (1985). Of the 523 species of living landsnails in Hubricht (1985), 146 species have Pleistocene occurrences, and 68 of these are extralimital species (i.e. species whose Pleistocene range limits are outside their present ones). We computed both the present day and Pleistocene latitudinal ranges of these species using the range maps in Hubricht (1985). Analyses of these ranges reveal a significant difference between the present day latitudinal ranges of extralimital and non-extralimital species; extralimital species tend to have a significantly shorter latitudinal range compared to the non-extralimital ones (Fig. 3). The pattern is thus opposite to that seen in the marine Pleistocene of California and in modern invasions.

Does the difference between the land snails and the marine mollusks simply reflect incomplete sampling in the case of the land snails, or does it result from the fundamental differences in the biology of the animals? Admittedly, sampling is far more complete for the marine mollusks of California compared to the Hubricht data, and better sampling of the Pleistocene land snail record is clearly needed. However, since geographically restricted species tend to have a lower probability of fossilization compared to widespread ones, the pattern seen in the latter case appears contrary to what would be predicted by poor paleontological sampling alone. Sampling aside, there are indeed a number of differences between the two systems, including the fact that in general land snails have much tighter habitat requirements compared to marine mollusks, and that in general land snail species appear to be more geographically restricted compared to their marine counterparts (Roy unpublished).

Another possibility is that in the case of the land snails, some of the observed differences reflect human impact. Present day geographic ranges of land snails have been heavily impacted by agriculture and the presence of introduced species. Hence for some species the difference between present and Pleistocene range limits may simply be the result of local extinctions due to human impact. This is difficult to quantify but archaeological records may provide some insight. A final possibility lies in the difference between the spatial scale of range shifts in the two systems. The analysis of Roy et al. (1995) focused on species that exhibit extensive range shifts, crossing major provincial boundaries and into different climatic and oceanographic regimes.

The land snails analyzed here, on the other hand, are characterized by much smaller range shifts, largely within a single biogeographic unit. This raises the intriguing possibility that the role of geographic range may be different for invasions that involve species shifting across major environmental barriers and into new climatic zones versus those involving more regional shifts.

What then is the role of geographic range in invasions, natural or otherwise? The above discussion shows how surprisingly little quantitative data exist on this, both for anthropogenic and Pleistocene cases. The handful

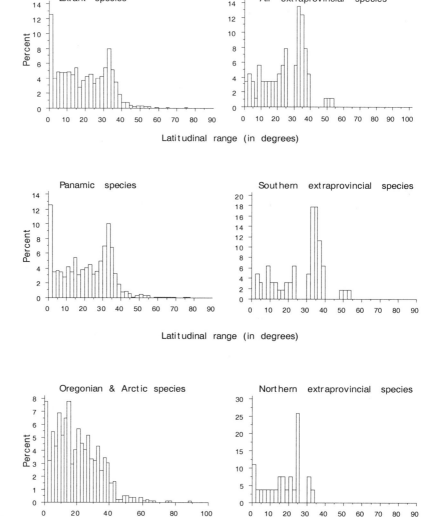

Figure 2. Comparisons of frequency distributions of present-day latitudinal ranges of eastern Pacific marine molluscan species. Each bar represents 2° of latitude. (A) Comparison of all extraprovincial species to the rest of the species pool; the frequency distributions are significantly different (p <0.0001 Mann-Whitney U test). (B) Comparison of the southern extraprovincial species to their parent species pool; distributions significantly different (p <0.0001 Mann-Whitney U test). (C) Comparison of northern extraprovincials to their parent species pool; distributions are not significantly different (p = 0.58 Mann-Whitney U test). (modified from Roy et al. 1995).

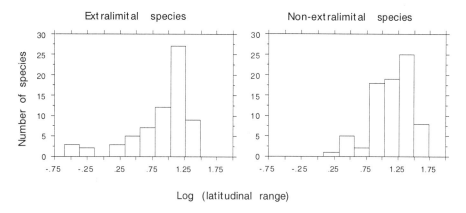

Figure 3. Present-day latitudinal ranges of extralimital and non-extralimital species of eastern North American native land snails. Latitudinal ranges compiled from Hubricht (1985). The distributions are significantly different (p = 0.0001 Mann-Whitney U test).

of studies quantitatively analyzing modern invasions (e.g. Moulton and Pimm 1986; Crawley 1987) have found large geographic range to be a good predictor of invasion success, and such an argument has also been made on the basis of qualitative evidence (e.g. Ehrlich 1989). The two Pleistocene datasets discussed here also support an important role of geographic range in invasions overall, but reveal additional complexities such as the difference between northern and southern extralimital species for marine mollusks, and differences between marine and terrestrial mollusks. In addition, Williamson (1996) argues that abundance rather than range may be the important predictor of invasion success and that the differences in geographic range simply reflect its correlation with abundance. Thus future analyses of the role of geographic range in invasions, both past and present, also need to take into account the effects of species abundance.

2.2 Taxonomic composition of invaders

Are successful invaders a random taxonomic subset of the species pool from which they are drawn, or are certain taxa over- or under-represented? For modern invasive plants on a global scale, statistical analyses suggest that successful invaders represent a non-random sample; certain plant families are significantly over-represented among the invaders (Daehler 1998; Pysek 1998). Similar taxonomic skew has also been documented for birds where the majority of the introduced species come from a few large families (Lockwood 1999). Many other lists of introduced species also suggest a strong taxonomic bias but appropriate statistical tests are lacking (e.g. see

North America and Britain, Williamson (1996) derived an average rate of spread of about 200 m per year. He concluded that the maximum rates could have approached 2 km per year for some taxa although most were about 600 m per year. These rates, remarkably high even for modern introductions, apparently characterized many native taxa although how they were achieved remains unclear (Bennett 1986; Williamson 1996). Similar rates calculations have not been attempted outside the pollen data although they should be possible for Quaternary insects (see Elias 1994) and may be possible even for some mammalian taxa (see FAUNMAP Working Group 1996). In any event the pollen example shows that high rates of spread are not unique to modern invasions but may be part of long term ecosystem dynamics.

3. HOMOGENIZATION PAST AND PRESENT

Shifts in species ranges in response to climatic change are an integral part of the long term dynamics of ecosystems. Such shifts in species ranges change local diversity as well as the differentiation between different areas, and the Pleistocene fossil record provide numerous examples of such changes (Roy et al. 1996; FAUNMAP Working Group 1996; Coope 1995; but see Jackson et al. 1996). For example, the Californian province in the late Pleistocene was home to a number of species of marine mollusks that now live only in tropical or cold temperate regions (Fig. 1). Geophysical and geochemical evidence now show that Pleistocene climatic fluctuations were much more dynamic than previously believed; climatic changes occurred at all temporal scales, from decades to centuries to millions of years, and ecological and paleoecological data reveal that species and communities respond to changes at all different scales (Roy et al. 1996). This dynamic picture of community composition preserved in the Pleistocene fossil record largely involves species that are still living, and provides insights into what the invasion process should look like in the absence of human impacts. Such baseline information is vital but is often hard to derive from neontological data on invasions because human impacts on many ecosystems predate such data (e.g. Jackson 1997). The challenge now is to quantify the differences between the patterns of anthropogenic and natural invasions and to understand what the long term consequences of these differences are.

Comparisons of ongoing invasions and their effects with those in the past are not without difficulties; the record is incomplete and spatial and temporal resolution decreases as one goes further back in time, and even the late Pleistocene examples may span a much longer timeframe compared to the ongoing invasions (Valentine and Jablonski 1992). On the other hand, it is encouraging that the resolution of the Pleistocene data is continuously increasing (Roy et al. 1996). More importantly statistical comparisons of general patterns are possible and should be attempted despite the differences in resolution. In the discussion above we highlighted aspects of the invasion process that are particularly suitable for such comparisons and that have important bearings on the question of homogenization. The issue of geographic range is particularly important not only in predicting invasion success but also in understanding the consequences of these invasions. The effect of invasions on the turnover in species composition between areas is partly dependent on whether the invasive species are widespread or geographically restricted. Most modern invasions apparently involve widespread species but the Pleistocene data suggest that under natural conditions, large range may not always be a good predictor of invasion success, and that the patterns may differ between taxa, between regions, and/or between land and sea. At present the evidence is admittedly tentative but it does suggest that the dominance of widespread species in modern invasions may, at least partially, reflect human influence.

The issue of taxonomic composition also remains seriously understudied. As discussed above there is a strong possibility that significant taxonomic skew may be the hallmark of anthropogenic influences on invasions and may not characterize most of the Pleistocene range shifts. Again, information is too scant at present to draw any firm conclusions but the issue deserves attention. The distribution of species within higher taxa varies from region to region, particularly with latitude, and these patterns reflect both biogeography and evolutionary history (Ricklefs and Schluter 1992; Gaston and Williams 1993; Williams et al. 1994; Roy et al. 1996). The introductions of exotic species change these taxonomic compositions that have evolved over evolutionary time, and such changes will clearly have a number of consequences including how species assemblages will respond to future global change (Vitousek et al. 1996). From a methodological perspective, such taxonomic skew along with the geographic range selectivity may also violate key assumptions of models that use species-area relationships to predict the future consequences of biotic mixing (see McKinney 1998).

At present we have very little data on natural rates of spread of taxa (Williamson 1996). Such baseline data are essential to put the current invasions into perspective, and to develop models that can predict how these invasions will change the spatial patterns of diversity and the taxonomic compositions of communities in the future. The Pleistocene and Holocene pollen record shows that the many plants can spread very rapidly in response to climate change. Animals may have responded in a similar manner as well, though the rates in the latter case are yet to be quantified. Exactly how the past rates for plants compare to ongoing invasions is uncertain but they appear to be high even by present day standards. Furthermore, future climate change will presumably trigger responses similar to those seen in the Pleistocene but if human-mediated introductions and habitat destructions change compositions of species assemblages in ways different from the past then predicting the consequences of future global change would indeed be difficult (Coope 1995).

4. SUMMARY

The ongoing patterns of biotic homogenization resulting from anthropogenic introductions of exotic species are unlikely to stop in the near future and the fossil record shows that all communities are invasible given the right circumstances (Valentine and Jablonski 1992). To what extent these ongoing invasions will homogenize regional and global biotas remains an open question. Comparisons of ongoing human-mediated invasions with those preserved in the Pleistocene and Holocene fossil record reveals many similarities but as discussed above, there also appear to be some crucial differences. These differences are not yet well quantified, but they deserve attention due to their potential for providing important clues about future consequences of biotic mixing. Such studies will nicely complement other approaches to the same problem that have successfully integrated neontological and paleontological data (Rosenzweig 1995; Brown and Lomolino 1998; McKinney 1998).

REFERENCES

Bennett, K. D. 1986. The rate of spread and population increase of forest trees during the postglacial. Philosophical Transactions of the Royal Society, London B 314:523-531.
Birks, H.J.B. 1989. Holocene isochrone maps and patterns of tree-spreading in the British Isles. Journal of Biogeography 16:503-40.
Brown, J. H. 1995. Macroecology. The University of Chicago Press, Chicago.
Brown, J. H. and M. V. Lomolino. 1998. Biogeography. Sinauer Associates, Massachusetts.
Carlton, J. T. and J. B. Geller. 1993. Ecological roulette: the global transport of nonindigenous marine organisms. Science 261:78-82.
Case, T. J. 1996. Global patterns in the establishment and distribution of exotic birds. Biological Conservation 78:69-96.
Coope, G. R. 1995. Insect faunas in ice age environments: why so little extinction? In: Extinction rates (J. H. Lawton and R. M. May, eds.), Oxford University Press, Oxford, Pp: 55-74.
Crawley, M.J. 1987. What makes a community invasible? Symposia of the British Ecological Society 26:429-53.
Daehler, C. C. 1998. The taxonomic distribution of invasive angiosperm plants: ecological insights and comparison to agricultural weeds. Biological Conservation 84:167-180.

Daehler, C. C. and D. R. Strong. 1993. Prediction and biological invasions. Trends in Ecology and Evolution 8:380.

Delcourt, H. R. and P. A. Delcourt. 1991. Quaternary ecology: A paleoecological perspective. Chapman and Hall, London.

Delcourt, P.A. and H.R. Delcourt. 1987. Long-term forest dynamics of the temperate zone. Springer-Verlag, New York.

di Castri, F. 1989. History of biological invasions with special emphasis on the Old World. In: Biological Invasions: a global perspective (J. A. Drake, H. A. Mooney, F. di Castri, R. H. Groves, F. J. Kruger, M. Rejmanek, and M. Williamson, eds.), John Wiley & Sons, Chichester, Pp: 1-30.

Elias, S. A. 1994. Quaternary insects and their environments. Smithsonian Institution Press, Washington.

Elrlich, P. R. 1989. Attributes of invaders and the invading process. In: Biological Invasions: a global perspective (J. A. Drake, H. A. Mooney, F. di Castri, R. H. Groves, F. J. Kruger, M. Rejmanek, and M. Williamson, eds.), John Wiley & Sons, Chichester, Pp: 315-328.

Elton, C. S. 1958. The ecology of invasions by animals and plants. Methuen, London.

FAUNMAP Working Group. 1996. Spatial response of mammals to Late Quaternary environmental fluctuations. Science 272:1601-1606.

Foote. M. 1994. Temporal variation in extinction risk and temporal scaling of extinction metrics. Paleobiology 20:424-444

Gaston, K. J. and P. H. Williams. 1993. Mapping the world's species – the higher taxon approach. Biodiversity Letters 1:2-8.

Gilpin, M. 1990. Ecological prediction. Science 248:88-89.

Grosholz, E. D. 1996. Contrasting rates of spread for introduced species in terrestrial and marine systems. Ecology 77:1680-1686.

Holway, D. A. 1998. Factors governing rate of invasion: a natural experiment using Argentine ants. Oecologia 115:206-212

Hubricht, L. 1985. The distributions of the native land mollusks of the Eastern United States. Fieldiana Zoology NS 24:1-191.

Jablonski, D. 1995. Extinctions in the fossil record. In: Extinction rates (J. H. Lawton and R. M. May, eds.), Oxford University Press, Oxford, Pp: 25-44.

Jackson, J. B. C. 1997. Reefs since Columbus. Coral Reefs 16 Suppl.:523-532.

Jackson, J. B. C, A. F. Budd, and J. M. Pandolfi. 1996. The shifting balance of natural communities? In: Evolutionary Paleobiology (D. Jablonski, D. H. Erwin, J. H. Lipps eds.), The University of Chicago Press, Chicago, Pp: 89-122.

Johnson L. E, and D. K. Padilla. 1996. Geographic spread of exotic species: Ecological lessons and opportunities from the invasion of the zebra mussel *Dreissena polymorpha*. Biological Conservation 78:23-33.

Lockwood, J. L. 1999. Using taxonomy to predict success among introduced avifauna: relative importance of transport and establishment. Conservation Biology 13:560-567.

Lodge, D. M. 1993. Biological invasions: Lessons for ecology. Trends in Ecology and Evolution 8:133-137.

Marshall, L. G., S. D. Webb, J. J. Sepkoski Jr., and D. M. Raup. 1982. Mammalian evolution and the Great American Interchange. Science 215:1351-1357.

McKinney, M. L. 1998. On predicting biotic homogenization: species-area patterns in marine biota. Global Ecology and Biogeography Letters 7:297-301.

McKinney, M. L. 1997. Extinction vulnerability and selectivity: Combining ecological and paleontological views. Annual Review of Ecology and Systematics 28:495-516.

Moulton, M.P. and S.L. Pimm. 1986. Species introductions to Hawaii. In: Biological Invasions of North America and Hawaii (H.A. Mooney and J.A. Drake, eds.), Springer-Verlag, New York, Pp: 231-49.

Petren, K. and T. J. Case. 1998. Habitat structure determines competition intensity and invasion success in gecko lizards. Proceedings of the National Academy of Sciences USA 95:11739-11744.

Pysek, P. 1998. Is there a taxonomic pattern to plant invasions? Oikos 82:282-294.

Reid, D. G. 1990. Trans-Arctic migration and speciation induced by climatic change: The biogeography of Littorina (Mollusca: Gastropoda). Bulletins of Marine Science 235:64-66.

Ricklefs, R. E. and D. Schluter. 1993. Species diversity: Regional and historical influences. In: Species diversity in ecological communities (R. E. Ricklefs and D. Schluter, eds.), The University of Chicago Press, Chicago, Pp: 350-363.

Rosenzweig, M. L. 1995. Species diversity in space and time. Cambridge University Press, Cambridge.

Roy, K., J. W. Valentine, D. Jablonski, and S. M. Kidwell. 1996. Scales of climatic variability and time averaging in Pleistocene biotas: implications for ecology and evolution. Trends in Ecology and Evolution 11:458-463.

Roy, K., D. Jablonski, and J. W. Valentine. 1995.Thermally anomalous assemblages revisited: Patterns in the extraprovincial range shifts of Pleistocene marine mollusks. Geology 23:1071-1074.

Sharov A. A., A. M. Liebhold, and E. A. Roberts. 1996. Spread of gypsy moth (Lepidoptera: Lymantriidae) in the central Appalachians: Comparison of population boundaries obtained from male moth capture, egg mass counts, and defoliation records. Environmental Entomology 25:783-792

Sheehan, P. M. 1988. Late Ordovician events and the terminal Ordovician extinction. New Mexico Bureau of Mines and mineral Resources Memoir 44:405-415.

Valentine, J. W., and D. Jablonski. 1993. Fossil communities: Compositional variation at many time scales. In: Species diversity in ecological communities (R. E. Ricklefs and D. Schluter, eds.),The University of Chicago Press, Chicago, Pp: 341-349.

Vermeij, G. J. 1991. When biotas meet: Understanding biotic interchange. Science 253:1099-1104.

Vermeij, G. J. 1989. Invasion and extinction: The last three million years of North Sea pelecypod history. Conservation Biology 3:274-281.

Vitousek, P. M., C. M. D'Antonio, L.L. Loope, and R. Westbrooks. 1996. Biological invasions as global environmental change. American Scientist 84:468-478.

Williams, P. H., C. J. Humphries, and K. J. Gaston. 1994. Centres of seed-plant diversity: the family way. Proceedings Royal Society, London B 256:67-70.

Williamson, M. H. 1996. Biological invasions. Chapman & Hall, London.

Chapter 3

Birds and Butterflies Along Urban Gradients in Two Ecoregions of the United States: Is Urbanization Creating a Homogeneous Fauna?

Robert B. Blair
Department of Zoology, Miami University, Oxford, Ohio 45056, USA

1.   INTRODUCTION

Humans are transforming Earth's landscape to an unprecedented degree in both magnitude and rate (Meyer and Turner 1992). We alter the landscape through the extraction of resources, the development of industry, the practice of agriculture, the pursuit of recreation, and the building of structures. These alterations affect every ecosystem on Earth (Vitousek et al. 1997).

An unintended consequence of this transformation is a phenomenon known as biological invasion -- the redistribution of species on Earth and the ecological and economic damages this incurs. Biological invasion is pervasive, alters ecosystem processes, and can reduce native biological diversity (Vitousek et al. 1996). This process is strongly linked with the transformation of the landscape in that human-altered ecosystems often provide the foci for invasion and -- in some cases -- the invaders hasten the transformation itself (Vitousek et al. 1997).

Much research on biological invasion has focused on the traits of invasive species, but several researchers have also identified specific features of landscapes that are invaded. A particular landscape is more likely to be invaded if it is similar in climate to the native habitat of the invading species, in an early successional stage, has low species diversity or low predator richness, and has been disturbed recently (See review by Lodge 1993). Many of these features are found in landscapes that are being urbanized (Blair 1996, Blair and Launer 1997, Blair 1999, Gering and Blair 1999).

Another unintended consequence of the transformation of Earth's landscape is widespread extinction. Wilcove et al. (1998) quantified the major threats to all taxa in the United States classified as imperiled by The Nature Conservancy or as endangered, threatened or candidates for listing under the Endangered Species Act. They found that habitat degradation and loss was the most common threat to all taxa combined as well as to birds and butterflies when these groups were examined individually. Alien species were the second most common threat. While this study includes many species that are not extinct by definition, it represents the many species that are on the path to extinction. As Hobbs and Mooney (1998) point out, extinction of a species is just the endpoint of a series of population extinctions. The major cause of many of these endangerments is transformation of the landscape.

The combination of these two unintended consequences of the transformation of the landscape -- invasion and extinction -- may result in the homogenization of Earth's biota (the theme of this book). Many researchers have suggested that human endeavors are leading to a more homogeneous fauna (Lodge 1993, Vitousek et al. 1997, Lodge et al. 1998, McKinney 1998), but few have examined if analogous communities in different ecoregions are actually becoming more homogeneous (i.e. similar) due to human transformation of the landscape. In this study, I examined birds and butterflies to test whether one type of human activity -- urbanization -- leads to more homogeneous biotic communities.

Specifically, I examined the idea that urbanization leads to the local extinction of species and invasion by other species and, consequently, creates more homogeneous faunal communities. I extended previous research on an urban gradient in California's coastal chaparral forest shrub ecoregion by comparing it to a similar gradient in Ohio's eastern broadleaf forest ecoregion (sensu Bailey et al. 1994). In California, I found that urbanization has predictable effects not only on the distribution and abundance of individual species of birds and butterflies, but also on community measures such as species richness, abundance, and Shannon diversity (Blair 1996, Blair and Launer 1997, Blair 1999).

In this study, I compared urban gradients in the coastal chaparral forest shrub ecoregion of California and the eastern broadleaf forest ecoregion of Ohio to test whether (1) increasing urbanization leads to increasingly homogeneous bird communities, (2) increasing urbanization leads to increasingly homogeneous butterfly communities, and (3) urbanization has the same homogenizing effects on these two taxa which have very different life histories. I also document the roles local extinction and invasion have in homogenizing these taxa.

2. METHODS

2.1 Field Sites

The sites chosen for this study represent forms of development typical in an urban-suburban matrix and comprise full gradients of human land-use ranging from relatively undisturbed to highly developed. They include a biological preserve, open-space reserve, golf course, residential neighborhood, office park (California) or apartment complexes (Ohio), and business district in each of the two ecoregions. The office park in California and the apartment complexes in Ohio are considered comparable sites due to the similarity in scale and layout of both buildings and plantings. I made this substitution because no office parks exist in the vicinity of the other sites in Ohio.

In California, the six sites were located within a 3 km radius circle centered at Stanford University near Palo Alto (Lat. 37°20' Long. 122° 15', typical elevation < 100 m). In Ohio, the six sites were located within a 10 km radius of Miami University, Oxford (Lat. 39°30' Long. 84° 45', typical elevation ~290 m). The sites within each ecoregion were judged to be ecologically similar prior to development according to historical maps and reference materials. Since urbanization is an exceedingly complex amalgamation of factors (McDonnell et al. 1993), the rank order of the sites from most natural to most urban initially was determined using the Delphi technique (See Blair 1996, Gering and Blair 1999).

I also used land cover as a surrogate measure of urbanization. At each study site, I calculated the area covered by buildings, pavement, lawn, grassland, and trees or shrubs in an approximately 50 m radius centered at each of 16 points used for the bird surveys (see below). I then converted the areas to percentage of site covered. In California, I estimated the cover of each land-use type from recent aerial photographs provided by Stanford University, the City of Palo Alto, and Jasper Ridge Biological Preserve. In Ohio, I used ortho-digitally corrected land-use maps from the City of Oxford and aerial photographs from the Department of Geography at Miami University.

Finally, I compared land cover on the urban gradients between the ecoregions. I did this with multivariate cluster analysis using Euclidean distance on the land-cover variables (percent cover buildings, pavement, lawn, grassland, and trees or shrubs) at all of the study sites. Euclidean distance is recommended for continuous data of equal scale (SYSTAT 1992).

2.2 Bird Survey Method

I estimated the densities of all perching or singing birds during peak breeding season using variable circular plots at 16 survey points within each site in an approximately 4 x 4 matrix where each point is at least 100 m from its nearest neighbor (Reynolds et. al 1980). In the preserve and open-space reserve in Ohio, sampling points were along foot trails instead of in a 4 x 4 matrix at the request of the areas' managers.

In California, points were surveyed a total of eight times in June and July 1992, and four times in June 1993. The open-space site was visited only six times in 1992 because it burned on 10 July. In Ohio, points were surveyed a total of eight times in June and July 1996, and four times in June 1997. Daily surveys began at dawn and continued until all 16 points at a site were covered, approximately two hours. Each point was visited for five minutes. This method results in an estimate of absolute density (birds ha^{-1}) of all species within each site (For details on method, see Blair 1996).

2.3 Butterfly and Skipper Survey Method

Butterflies (Papilionoidea) and skippers (Hesperioidea) (collectively referred to as "butterflies") were sampled along walking transects. I combined two bird-survey points into one butterfly-sampling area of approximately 2 ha. This created 8 areas within each land-use type for a total of 48 sampling areas across each urban gradient. These areas were walked twice (a.m. and p.m. on different days) for 15 minutes for each of four sampling rounds in California and four times (twice in a.m. and twice in p.m. on different days) for each of two sampling rounds in Ohio. All species seen during each round were noted and voucher specimens of species difficult to identify were captured. This method provides a measure of relative abundance based on the frequency of observation of each species, which was then converted to a percentage. Abundance may range from 0 (no individuals of that species seen during any sampling period at that site) to 100 (at least one individual of that species seen during each 15-minute sampling period at that site) [See Blair and Launer (1997) for details].

In California, butterflies were sampled during four periods: 24 Aug. - 1 Sep. 1992, 22 Mar. - 19 Apr. 1993, 25 May - 9 June 1993, and 13 Aug. - 18 Aug. 1993. These are periods of peak flight phenology in this region (Opler and Langston 1968). In Ohio, butterflies were sampled during two periods: 8 Jul. - 1 Aug. 1997, and 2 Jul. - 24 Jul. 1998, also periods of peak flight phenology (Iftner et al. 1992).

2.4　Homogenization of Fauna

I used two methods to assess if increasing urbanization led to a more homogeneous fauna. First, for birds alone and butterflies alone, I compared sites within land-use type between ecoregions (e.g. golf courses in both ecoregions) using Jaccard's index of similarity, which measures percent species overlap between two sites (Magurran 1988). Jaccard's Index may range from 0 (indicating no species overlap) to 1 (indicating complete species overlap). Second, for each taxon, I performed a multivariate cluster analysis of all species in all of the 12 formal study sites. I used normalized percent disagreement on the average daily density for birds or the frequency of observation for butterflies. Normalized percent disagreement is a metric that is useful when there is high ß-diversity and, consequently, many empty cells (Jongman et al. 1995, SYSTAT 1992). The cluster analysis is a more comprehensive assessment of homogenization because it compares each site to all others and 'recognizes' subtle shifts in distribution. To illustrate, a species might be present only in the business district in California and only in the apartment complexes in Ohio. In this case, Jaccard's index comparing California and Ohio would be zero because there is no species overlap within the same land-use type. However, the cluster analysis would indicate some level of similarity because it compares within land-use types as well as between them and would group the business district and apartment complexes.

2.5　Comparing Taxa and Effects of Homogenization

To assess if urbanization has a similar effect on homogenization of birds and butterflies, I compared the Jaccard's indices generated during the comparison of ecoregions (above) for each taxon using Pearson correlations. I also visually compared the cluster analyses developed for each taxa.

2.6　Local Extinction and Invasion Along the Gradients

The best representatives of the communities that existed in each ecoregion prior to European settlement are assumed to be those species found in the preserves. This is otherwise known as the minimally-disturbed standard (Karr and Chu 1998). Local extinction along the gradient is assumed to have taken place if a species is not found at more urbanized sites. Local invasion is assumed to have taken place if a species is only found at more urbanized sites along the gradient. These definitions are not mutually exclusive. For example, House Finches do not occur at either end of the gradient in California but do occur at sites of intermediate levels of urbanization. Consequently, this species invades with low levels of urbanization but then goes extinct with higher levels.

3. RESULTS

3.1 Assessment of the Urban Gradient

The Delphi technique led to similar orderings of the sites along the gradient in both ecoregions. In California, the ranking of the sites in order of increasing urbanization was biological preserve, open-space reserve, golf course, residential area, office park, and business district. In Ohio, the ranking of the sites was identical except that exactly half of the respondents ranked the apartment complexes and half ranked the business district as most urban. Land cover data (Figure 1) suggest that the business district is the more urban site (Blair 1996, Gering and Blair 1999).

The assessments of land cover revealed unimodal distributions of percentage area covered by buildings, pavement, lawn, grassland, and trees and shrubs along the urbanization gradient (Figure 1). Cluster analysis using these measures generally grouped each land-use type across ecoregions (i.e. the residential areas in California and Ohio were more similar to one another according to cover characteristics than to other sites) (Figure 2). The major exception was the open-space reserve in California. This location had large areas of grassland, a cover type largely unique to that site. The distinct placement of the open-space reserve in California forced the open-space reserve in Ohio to group with the preserves.

3.2 Birds Along the Gradient

The densities of all bird species varied to some degree across the gradients (Figures 3 and 4). Of the 40 bird species in the California survey, 31 had continuous, unimodal distributions (within the limits of one standard error of their estimated daily densities) of abundance across the urban gradient. In other words, the estimated densities of these species were highest at one site along the gradient and decreased gradually as urbanization increased or decreased. Of the 44 species encountered in Ohio, 38 had unimodal distributions within the limits of one standard error.

3.3 Butterflies Along the Gradient

The densities of all butterfly species varied to some degree across the gradients (Figures 5 and 6). Of the 26 butterfly species in the California

3. Birds and Butterflies Along Urban Gradients

survey, 21 had continuous distributions across the urban gradient, and 5 had disjunct distributions. In other words, most species were found in single sites or in blocks of two or three contiguous sites along the gradient. Of the 28 butterfly species in the Ohio survey, 22 had continuous distributions across the urban gradient, and 6 had disjunct distributions.

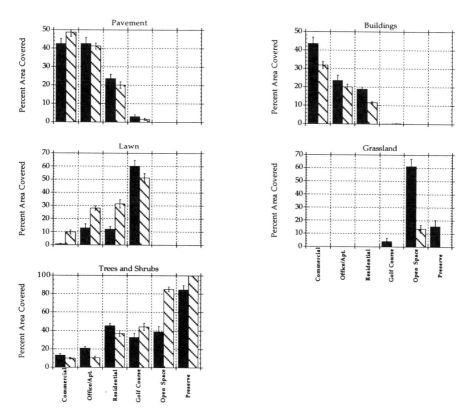

Figure 1. Percent area (± s.e. n = 16 points) covered by pavement, building, lawn, grassland, and trees and shrubs in California (darker) and Ohio (lighter).

3.4 Homogenization of Birds

The Jaccard's Index values for species similarities for birds between the sites in California and Ohio were 0.15 for the business districts, 0.22 for the office park/apartments, 0.15 for the residential areas, 0.11 for the golf courses, 0.03 for the open-space reserves, and 0.06 for the biological preserves (Figure 7). These results suggest that the land-use types fall into two groups: relatively urbanized sites (the business district, office parks/apartments, residential areas, and golf courses) with an average

Jaccard's Index of 0.16 and relatively undeveloped sites (the open-space reserve and biological preserve) with an average Jaccard's of 0.05. Instead of increasingly monotonically with urbanization, it appears that species similarity crosses a threshold with a substantial increase between the open-space reserves and the golf courses -- a point where the landscape becomes entirely manipulated.

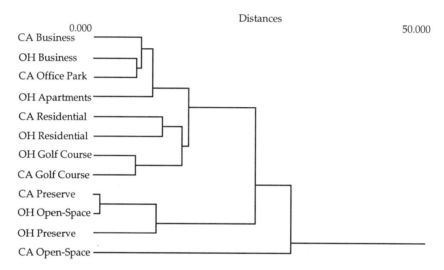

Figure 2. Cluster analysis of sites in California and Ohio based on percent area covered by pavement, building, lawn, grassland, trees and shrubs and the metric of Euclidean distance.

The cluster analysis revealed a similar pattern (Figure 8). Based on the average daily density of all bird species, the dendrogram shows that the biological preserves, open-space reserves, and golf courses contain relatively distinct assemblages, whereas the residential areas, office park/apartment complexes, and business districts fall into a single cluster regardless of ecoregion.

3.5 Homogenization of Butterflies

The Jaccard's Index values for butterfly species between the sites in California and Ohio were 0.10 for the business districts, 0.15 for the office

3. Birds and Butterflies Along Urban Gradients

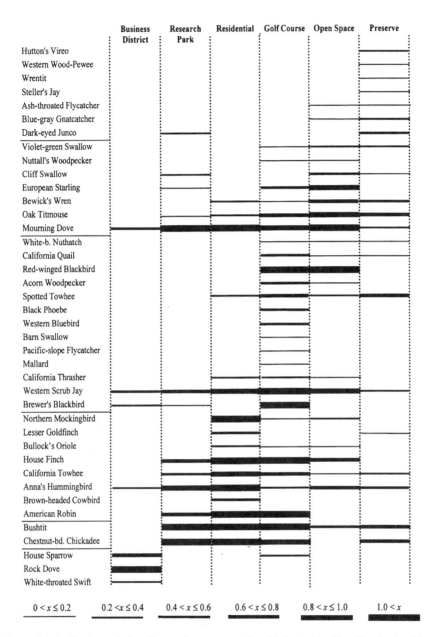

Figure 3. Distribution and abundance of summer resident birds in Palo Alto, California. Line width represents birds per hectare. Species are arranged in order of greatest density along the urban gradient. Lines in left column represent breaks between land-use types.

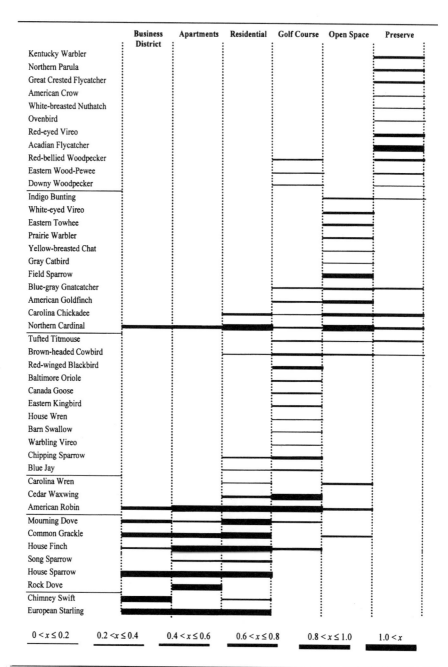

Figure 4. Distribution and abundance of summer resident birds in Oxford, Ohio. Line width represents birds per hectare. Species are arranged in order of greatest density along the urban gradient. Lines in left column represent breaks between land-use types.

3. Birds and Butterflies Along Urban Gradients

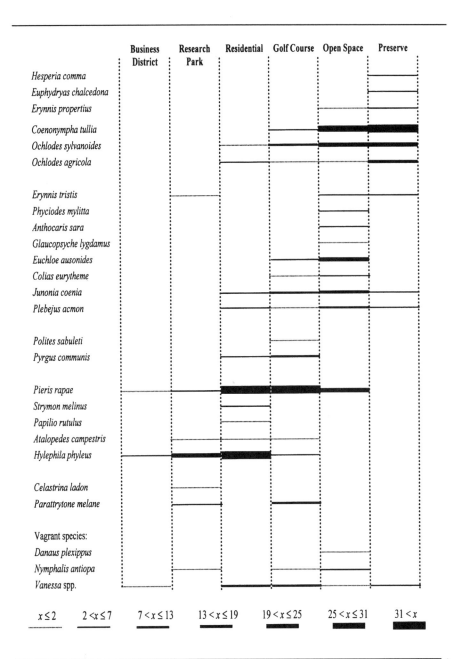

Figure 5. Distribution and abundance butterflies in Palo Alto, California. Line width represents relative frequency of observation. Species are arranged in order of greatest abundance along the urban gradient.

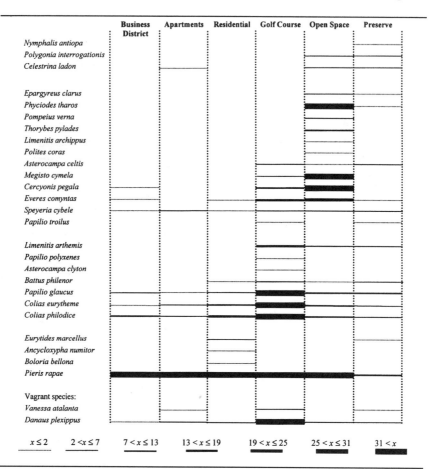

Figure 6. Distribution and abundance butterflies in Oxford, Ohio. Line width represents relative frequency of observation. Species are arranged in order of greatest abundance along the urban gradient.

park/apartments, 0.05 for the residential areas, 0.11 for the golf courses, 0.12 for the open-space reserves, and 0.04 for the biological preserves (Figure 7). These results suggest that urbanization has not increased the homogeneity of butterfly faunas between the two ecoregions. The butterfly fauna of the two regions do not exhibit a high degree of overlap. Only four species -- *Celastrina ladon, Colias eurytheme, Pieris rapae,* and *Danaus plexippus* -- were documented in the surveys of both ecoregions. Despite this paucity of common butterfly species, the cluster analysis reveals a pattern similar to that of the bird fauna (Figure 8). Based on the frequency of observation of all butterfly species in all sites, the dendrogram shows that the biological preserves, open-space reserves, and golf courses are fairly distinct assemblages, whereas the residential areas, office park/apartment complexes, and business districts fall into one cluster regardless of ecoregion.

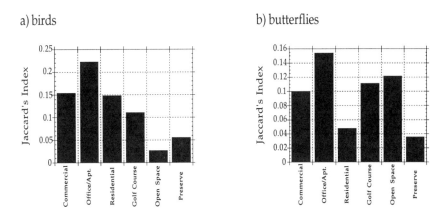

Figure 7. Jaccard's Index of similarity for a) birds and b) butterflies comparing land-use sites in California and Ohio. The numbers for birds and butterflies are not significantly correlated (n = 6, p = 0.458, R^2 = 0.144).

3.6 Comparing Taxa and Effects of Homogenization

Urbanization appears to have a greater effect on the homogenization of birds than butterflies. Of the 40 species of birds in California and 44 species in Ohio, 11 species were common to both ecoregions. Of the 26 species of butterflies in California and 28 species in Ohio, only 4 species were common to both ecoregions. There was no significant correlation between the Jaccard's Index values for the birds and butterflies (n = 6, p = 0.458, R^2 = 0.144).

The cluster analyses of both birds and butterflies suggest that some degree of homogenization of both taxa occurs with urbanization. Both cluster analyses have the less urbanized sites (the preserves, open-space reserves, and golf courses) containing relatively distinct assemblages of species. In contrast, the most urbanized sites (the business districts, office parks/apartments, and residential areas) fall out in a single cluster regardless of ecoregion, which implies overlapping assemblages of species.

3.7 Local Extinction and Invasion Along the Gradients

A strong pattern of extinction of the woodland species -- the minimally disturbed standard -- exists for both birds and butterflies in both ecoregions. Most of these species gradually dropped out along the gradient as the sites became more urbanized. Of the 21 bird species present in the preserve in California, only three were also present in the business district. Of the 17 bird species present in the preserve in Ohio, only one was also present in the business district (Table 1). The butterflies in both ecoregions exhibited the same pattern. Of the 10 butterfly species present in the preserve in California, none were present in the business district. Of the 18 butterfly species present in the preserve in Ohio, seven were present in the business district (Table 2). These patterns were also highly nested -- once a woodland species dropped out at a less urbanized site, it rarely reappeared at a more urbanized one along the gradient.

The pattern of invasion along the gradients is strong as well for both birds and butterflies in both ecoregions. In California, 19 species of birds and 16 species of butterflies can be considered local invaders because they were not present at the preserve but were present at more urbanized sites. In Ohio, 27 species of birds and 10 species of butterflies can be considered local invaders. Being a local invader did not necessarily make these species safe from local extinction due to urbanization, however. Of these subsets of invaders, 14 species of birds and 14 species of butterflies in California, and 20 species of birds and 9 species of butterflies in Ohio, did not occur in the most urbanized sites. In other words, these species were able to exploit suburban conditions but were not able to tolerate the most urban ones. These patterns can be seen in Figures 3 through 6 by noticing the shift in species densities as your eye travels down each figure.

4. DISCUSSION

"For the first time in the history of the human species, two changes are now impending. One is the exhaustion of the wilderness in the more habitable portions of the globe. The other is the world-wide hybridization of cultures through modern transport and industrialization." (Leopold 1949).

Aldo Leopold (1949) made this observation concerning the world-wide hybridization of cultures more than fifty years ago. Unfortunately, it is still the lament of anthropologists, linguists, social scientists, policymakers, and ecologists. If Leopold were alive today, he could easily extend his argument and conclude that the biological results of this cultural hybridization are the dual phenomena of landscape transformation and biological invasion.

The transformation of the landscape occurs when humans settle an area and then convert the land to "productive" human enterprises such as agriculture, industry, and recreation (Vitousek et al. 1997). The desire and ability to transform the land leads to the homogenization of the landscape -- a phenomenon that is just now being recognized (Clay 1994). With this

3. Birds and Butterflies Along Urban Gradients

homogenization, one could argue that the outskirts of any given town have begun to resemble the outskirts of any other town. This homogenization not only affects the physical condition of the land, but also the flora and fauna that inhabit it.

4.1 Urbanization and a Homogeneous Fauna

This study addresses one specific type of landscape conversion -- urbanization -- in two distinct ecoregions: California's coastal chaparral forest shrub and Ohio's eastern broadleaf forest. The differences between these two ecoregions -- geographic, environmental, and economic -- would suggest that no amount of manipulation would lead to a similar fauna. The results of this study demonstrate that urbanization does homogenize fauna.

The coastal chaparral forest shrub represents stereotypical California. It has a Mediterranean climate with a rainy season between November and April and virtually no precipitation between May and October. Its temperatures rarely dip below freezing or rise above 30° C. People can grow almost anything (with the aid of irrigation) which leads to an almost surreal flora of eastern hemlock and palm trees, pin oaks and coast redwoods. The region is densely populated and Palo Alto is located in the heart of the Silicon Valley, a megalopolis that extends 100 km from San Francisco to San Jose. The area has experienced an unprecedented building boom since the early 1970s.

The eastern broadleaf forest, in contrast, represents stereotypical Midwestern United States. It experiences moderate precipitation throughout the year, with the highest levels in May and June. Temperatures frequently remain below freezing in winter, and often rise above 30° C in summer. Humans are limited to growing annuals and plants that are freeze tolerant, though they do not need to actively irrigate most species. The region is predominantly rural and Oxford is separated by more than 50 km from any major urban area. It has not experienced a building boom and its population has remained relatively stable at 8,000 permanent residents and 16,000 students since the early 1970s.

Though these regions are very different climatically and economically, both the birds and butterflies within each ecoregion respond similarly to urbanization at the community level with respect to such measures as species richness, Shannon diversity, and abundance (Blair in review). When particular species are taken in account, it becomes apparent that the bird and butterfly communities become more similar among ecoregions as the sites become more urban.

The bird communities display the clearest patterns of the homogenization that accompanies urbanization. A land-use type by land-use type (e.g. golf course to golf course) comparison using Jaccard's Index shows that the most natural sites (the preserves and open-space reserves) in California and Ohio had an average overlap in bird species of approximately 5 percent. In contrast, the most developed sites (the business districts, office parks/apartments, residential areas, and golf courses) had an average overlap in species of approximately 16 percent -- more than three times the level of the most natural sites.

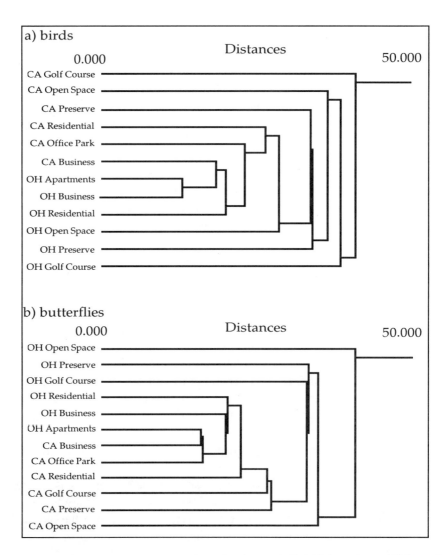

Figure 8. Cluster analysis of sites based on abundance of all a) bird species and b) butterfly species using the distance metric of normalized percent disagreement. The office park in California and the apartments in Ohio are considered equivalent land uses in this study.

3. Birds and Butterflies Along Urban Gradients

Table 1. Loss of the original oak-woodland bird species across the urban gradient in California and Ohio. Only those species found at the biological preserves are listed. A (+) indicates presence.

Bird Species	Preserve	Open Space	Golf Course	Residential	Office/ Apts.	Business
California Ecoregion						
Hutton's Vireo	+					
Western Wood-Pewee	+					
Steller's Jay	+					
Wrentit	+					
Dark-eyed Junco	+				+	
Ash-throated Flycatcher	+	+				
Blue-gray Gnatcatcher	+	+			+	
Cliff Swallow	+	+				
White-breasted Nuthatch	+	+	+			
California Quail	+	+	+			
Violet-green Swallow	+	+	+			
Spotted Towhee	+	+	+	+		
Bewick's Wren	+	+	+	+		
Lesser Goldfinch	+			+	+	
Oak Titmouse	+	+	+	+	+	
California Towhee	+	+	+	+	+	
Bushtit	+	+	+	+	+	
Chestnut-bd. Chickadee	+		+	+	+	
Western Scrub Jay	+	+	+	+	+	+
Anna's Hummingbird	+	+	+	+	+	+
Mourning Dove	+	+	+	+	+	+
Ohio Ecoregion						
Acadian Flycatcher	+					
American Crow	+					
Great Crested Flycatcher	+					
Kentucky Warbler	+					
Red-eyed Vireo	+					
Ovenbird	+					
White-breasted Nuthatch	+					
Northern Parula	+					
Indigo Bunting	+	+				
Tufted Titmouse	+	+	+			
Blue-gray Gnatcatcher	+	+	+			
Red-bellied Woodpecker	+		+			
Downy Woodpecker	+		+			
Eastern Wood-Pewee	+		+			
Brown-headed Cowbird	+	+	+	+		
Carolina Chickadee	+	+	+	+		
Northern Cardinal	+	+	+	+	+	+

Table 2. Loss of the original oak-woodland butterfly species across the urban gradient in California and Ohio. Only those species found at the biological preserves are listed. A (+) indicates presence.

Butterfly Species	Preserve	Open Space	Golf Course	Residential	Office/ Apts.	Business
California Ecoregion						
Hesparia comma	+					
Euphydryas chalcedona	+					
Paratrytone melane	+				+	
Erynnis tristis	+	+			+	
Erynnis propertius	+	+				
Coenonympha tullia	+	+	+			
Ochlodes sylvanoides	+	+	+	+		
Ochlodes agricola	+	+	+	+		
Junonia coenia	+	+	+	+		
Plebejus acmon	+	+	+	+		
Ohio Ecoregion						
Nymphalis antiopa	+					
Papilio troilus	+		+			
Eurytides marcellus	+			+		
Polygonia interrogationis	+	+				
Epargyreus clarus	+	+				
Phyciodes tharos	+	+				
Celestrina ladon	+	+			+	
Asterocampa celtis	+	+	+			
Limenitis arthemis	+	+	+			
Battus philenor	+	+	+	+		
Vanessa atalanta	+		+		+	
Everes comyntas	+	+	+	+	+	+
Speyeria cybele	+	+	+	+	+	+
Colias eurytheme	+	+	+	+	+	+
Colias philodice	+	+	+	+	+	+
Papilio glaucus	+	+	+	+	+	+
Pieris rapae	+	+	+	+	+	+
Danaus plexippus	+	+	+	+	+	+

The more nuanced, multivariate cluster analysis, which compares the density of all bird species in all sites to all others in both ecoregions, reveals that the most natural sites along the gradient (the preserves, open-space reserves, and golf courses) are relatively distinct. This implies that the avifauna of these three land-use types is fairly distinct to either Ohio or California. In contrast, those sites that are more developed (the business districts, office park/apartments, and residential areas) cluster together regardless of ecoregion. In other words, the assemblages of bird species are relatively distinct in natural sites but many of the same species are found in urban sites irrespective of ecoregion.

The butterfly communities do not display as clear a pattern concerning homogenization and urbanization as do the bird communities. The land-use type by land-use type comparison shows no clear trend between community similarity and urbanization, nor does it show a distinct break as the sites become more urban. However, the multivariate cluster analysis, which compares all sites to all others, results in an almost identical pattern to that of the birds. Namely, the most natural sites along the gradient (the preserves, open-space reserves, and golf courses) divide along ecoregion while the more developed sites (the business districts, office park/apartments, and residential areas) fall out in a single cluster regardless of ecoregion. This indicates that, even for butterflies, the sites become at least marginally more similar as they become more urban.

These comparisons suggest that the more humans manipulate the environment, the more similar the faunal communities become. However, they also suggest that this phenomenon is more pronounced with birds than with butterflies.

The levels of species similarity between the two ecoregions appear to be rather low for both the birds and butterflies -- the maximum overlap being approximately 22 percent for the birds in the office park/apartment complexes. This raises the question of how much these two communities overlap in total. Forty species of birds were regularly present in California and 44 species were regularly present in Ohio with only 11 species common to both ecoregions. This creates a total avian species overlap (comparing all species in all sites) between the ecoregions of approximately 15.1 percent. In contrast, 26 species of butterflies were present in California and 28 species in Ohio with only 4 species common to both and a total butterfly species overlap of approximately 8.0 percent. The lower level of overlap in butterfly species may explain, in part, the stronger pattern in homogenization for birds as compared to butterflies.

The low level of overlap in species points to regional differences in the entire fauna of both birds and butterflies. For example, White-throated Swifts are found in downtown Palo Alto while Chimney Swifts are found in uptown Oxford. They both occupy a similar role in the environment but they cannot be directly compared with indices of species overlap. Consequently, a comparison of the two ecoregions could not result in identical bird or butterfly communities with high levels of urbanization. Rather, this points out that urbanization leads to more similar communities. This similarity -- but not identity -- is presumably due to a combination of environmental and dispersal limitations on many species.

4.2 The Difference Between Birds and Butterflies

The subtly different response by birds and butterflies to urbanization raises the question of how these taxa differ and what may bring about this difference. The answer may lie in the life history characteristics of the taxa and each taxon's ability to invade. The generic characteristics of invading plant and animal species that are most cited in the invasion literature include large native ranges, high abundance, high rates of dispersal, polyphagy, short generation times, high genetic variability, single-parent reproduction, phenotypic plasticity, eurytopy, and human commensalism. However, predicting whether a specific organism will be a successful invader from these traits is an unreliable science (Lodge 1993, Lockwood 1999).

Little theoretical work has considered whether birds or butterflies would make better invaders based on these characteristics. However, empirical data from California suggest that birds, on the whole, are better invaders. Of 284 species of birds in California, 19 (6.7 percent) are introduced. Of 254 species of butterflies, only 1 (0.4 percent) is introduced (Data from Hobbs and Mooney 1998). Additionally, many researchers have addressed biological invasion by birds (e.g. Lockwood 1999, Case 1996) but few, if any, have addressed invasion by butterflies except in very applied situations such as that of the gypsy moth (Gerardi and Grimm 1979).

This difference in the ability to invade may be related to the highly co-evolved relationships between butterflies and their host plants (Ehrlich and Raven 1964). Butterflies often rely on one or a limited number of species of host plants at three different stages in their development: as vegetation-eating larvae, as nectar-feeding adults, and as ovipositing adults (generally ovipositing on the same plant that larvae use). This means that for a butterfly to successfully invade new habitat, it must either switch host plants (an unlikely and usually lethal event) or be introduced after its host plants have already established themselves in the new habitat (i.e. a sequential double invasion). This idea is supported by the fact that the butterfly species that has invaded California (mentioned above) is *Pieris rapae* -- the cabbage butterfly -- and it has an unusually wide host plant diet that includes most of the Brassicaceae and all of the cultivated cruciferous vegetables (Garth and Tilden 1986). This suggests that the cabbage butterfly invaded after humans started cultivating its host plants in North America.

4.3 A Possible Mechanism: Extinction plus Invasion

The patterns exhibited by both birds and butterflies on the gradients in this study suggest that local extinction and local invasion regularly occur with urbanization. Woodland species gradually disappear along the gradient as the sites become more urbanized. Assuming that these species were evenly distributed across the landscape prior to European settlement, this implies that they have gone 'extinct' with urbanization. The use of the term 'extinct' is not traditional in the sense that the entire species has disappeared, rather I use it in the sense that a population has become locally extirpated. Hobbs and Mooney (1998) highlighted that population extinction is what should be of concern to conservation biologists because species extinction is just the final result of myriad population extinctions.

Local invasion is also a dominant pattern in both birds and butterflies along these gradients. Over 47 percent of the bird species in California and 61 percent in Ohio did not occur in the most natural sites, which implies that they may be considered local invaders that expanded their populations with European settlement in the area. The same held true for the butterfly communities since 61 percent of the species in California and 35 percent in Ohio did not occur in the most natural areas. These percentages suggest that a substantial percentage of species have been able to take advantage of the changes in the environment that urbanization brings about. However, it does not imply that these species, themselves, are not subject to the effects of more intense urbanization. Many of the species that were able to invade locally at moderate levels of urbanization also went locally extinct with more intense urbanization. For example, the butterfly -- *Megisto cymele* -- was not present in the preserve in Ohio, did occur in the open-space reserve and the golf course, and was not present in any of the more urbanized sites. Consequently, it was both an invader that capitalized on some transformation of the landscape, and a species that went locally extinct with more drastic levels of urbanization (Figure 6).

Numerous mechanisms resulting from urbanization may contribute to the extinction, invasion, and consequent homogenization documented in this study. The transformation of the landscape by urbanization does not occur by changing one environmental factor such as pH or one ecological process such as extinction. Rather, urbanization occurs in a patchy network of many environmental parameters and processes that play out at many different scales. At the landscape level, urbanization changes patch size, configuration, connectivity, the amount of edge, and ecological processes (Vitousek et al. 1996). For example, predation rates on artificial nests in Ohio decline monotonically as the sites become more urban (Gering and Blair 1999). At the population level, urbanization may affect the viability and longevity of isolated populations (Bolger et al. 1991). At the territory level, urbanization may effect physical environmental parameters and the dispersion of resources. Here, I suggest that one of the effects of urbanization is how it alters local extinction and invasion.

4.4 Research Needs

The intersection of transformation of the landscape, extinction, and invasion can be explored further in many ways. First, an examination of historical records would be useful to see how tightly urbanization and extinction are linked. Is extinction an immediate process? Does it have a predictable lag period as suggested by Bolger et al. (1991)? Does extinction precede invasion? Case (1996) suggests that extinction is a good predictor of invasion because it represents previous transformation of the landscape. This may be contrary to the idea that invasion by some species directly leads to extinction of others. Rather, it implies that it may be *either* a direct association, such as when non-native fish crowd out native species (Williams et al. 1989), *or* a more serendipitous association such as when a recovering clearcut provides the conditions for edge species to increase in number and diminishes the numbers of interior species. Another avenue of research would be to review the history of invasion by butterfly species. How often has it occurred in comparison to other taxa? Under what conditions does it take place? What are the characteristics of an invasive butterfly? Finally, the idea that homogenization is the result of extinction and invasion should be examined in the context of global change (See McKinney, this volume). What patterns that are occurring globally -- such as the northward expansion of butterflies in Europe (Parmesan et al. 1999) -- may be attributed to human transformation of Earth?

5. CONCLUSION

With this study, I evaluate the effects of urbanization on bird and butterfly communities by comparing species distributions on urban gradients in two ecoregions of the United States: the coastal chaparral forest shrub ecoregion of California and the eastern broadleaf forest ecoregion of Ohio. First, I tested whether increasing urbanization leads to increasingly homogeneous bird communities between the two ecoregions. The answer is yes. Increasing urbanization in the two ecoregions leads to a more similar -- but not identical -- bird fauna. This increase in similarity does not appear to occur gradually, but rather in a step function when the landscape is transformed from not maintained to highly maintained. Second, I tested whether increasing urbanization leads to increasingly homogeneous butterfly communities. The answer is a qualified yes. Few species of butterflies overlap between the two ecoregions but these surveys show that the butterfly communities do become more similar -- at least to a small degree. Third, I tested whether urbanization has the same homogenizing effects on these two taxa that have very different life histories. The answer is no, the effect differs in magnitude. The homogenizing effect is stronger on the bird community, most likely because birds are more flexible invaders that don't rely on certain species of host plants to the degree that butterflies do. Finally, I documented the roles that local extinction and local invasion play in homogenizing these taxa. Both taxa display strong patterns of local extinction and local invasion under the stress of urbanization.

ACKNOWLEDGEMENTS

A number of people have helped me in this project. Alistair Hobday, Charlie Quinn, Russell Bell, Valerie Tierce, and Julie Whipkey deserve thanks for help in the field. Paul Ehrlich, Alan Launer, Erica Fleishman, Mike Vanni, Dave Berg, and Ann Rypstra deserve thanks for helping me assemble my thoughts on this topic. Finally, Erica Fleishman, Dave Berg, Michael McKinney, and an anonymous reviewer helped with the execution of this paper.

REFERENCES

Bailey, R. G., P. E. Avers, T. King, and W. H. McNab, eds. 1994. Ecoregions and subregions of the United States (map). Washington, DC; U.S. Geological Survey. Scale 1:7,500,000; colored. Accompanied by a supplementary table of map unit descriptions compiled and edited by McNab, W. H., and R. G. Bailey, Prepared for the U.S. Department of Agriculture, Forest Service.

Blair, R. B. In Review. Creating a homogeneous avifauna: Local extinction and invasion along urban gradients in California and Ohio.

Blair, R. B. 1999. Birds and butterflies: surrogate taxa for assessing biodiversity? Ecological Applications 9(1):164-170.

Blair, R. B. 1996. Land use and avian species diversity along an urban gradient. Ecological Applications 6(2):506-519.

Blair, R. B. and A. E. Launer. 1997. Butterfly diversity and human land use: species assemblages along an urban gradient. Biological Conservation 80:113-125.

Bolger, D. T., A. C. Alberts and M. E. Soulé. 1991. Occurrence patterns of bird species in habitat fragments: sampling, extinction, and nested species subsets. The American Naturalist 137(2):155-166.

Case, T. J. 1996. Global patterns in the establishment and distribution of exotic birds. Biological Conservation 78:69-96.

Clay, G. 1994. Real places: an unconventional guide to America's generic landscape. The University of Chicago Press, Chicago, Illinois.

Ehrlich, P. R. and P. H. Raven. 1964. Butterflies and plants: a study in coevolution. Evolution 18:586-608.

Garth, J.S. and J.W. Tilden. 1986. California Butterflies. University of California Press, Berkeley, California. 246 pp.

Gerardi, M. H. and J. K. Grimm 1979. The history, biology, damage, and control of gypsy moth, *Porthetria dispar* (L.). Fairleigh Dickinson University Press, Rutherford, New Jersey. 233 pp.

Gering, J. C. and R. B. Blair. 1999. Predation on artificial bird nests along an urban gradient: predatory risk or relaxation in urban environments? Ecography 22:532-541.

Hobbs, R. J. and H. A. Mooney. 1998. Broadening the extinction debate: Population deletions and additions in California and Western Australia. Conservation Biology 12(2):271-283.

Iftner, D. C., J. A. Shuey, and J. V. Calhoun. 1992. Butterflies and Skippers of Ohio. Ohio Biological Survey Bulletin New Series Vol. 9 No. 1 xii + 212 pp.

Jongman, R. H. G., C. J. F. Ter Braak, and O. F. R. Van Tongeren. 1995. Data Analysis in Community and Landscape Ecology. Cambridge University Press, Cambridge 299 pp.

Karr, J. R. and E. W. Chu. 1998. Restoring life in running waters: Using multimetric indexes effectively. Island Press, Covelo, California.

Leopold, A. 1949. A Sand County almanac and sketches here and there. Oxford University Press, New York. 226 pp.

Lodge, D. M., R. A. Stein. K. M. Brown, A. P. Covich, C. Brönmark, J. E. Garvey and S. P. Klosiewski. 1998. Predicting impact of freshwater exotic species on biodiversity: Challenges in spatial scaling. Australian Journal of Ecology 23:53-67.

Lodge, D. M. 1993. Biological invasions: Lessons for ecology. Trends in Evolution and Ecology 8(4)133-137.

Lockwood, J. L. 1999. Using taxonomy to predict success among introduced avifauna: Relative importance of transport and establishment. Biological Conservation 13(3):560-567.

Magurran, A. E. 1988. Ecological diversity and its measurement. Princeton University Press, Princeton, New Jersey. 179 pp.

McDonnell, M. J., S. T. A. Pickett, and R. V. Pouyat. 1993. The application of the ecological gradient paradigm to the study of urban effects. In: Humans as components of ecosystems (M. J. McDonnell and S. T. A. Pickett, eds.), Springer-Verlag, New York, Pp: 175 – 189.

McKinney, M. L. 1998. On predicting biotic homogenization: species-area patterns in marine biota. Global Ecology and Biogeography Letters 7:297-301.

Meyer, W. B. and B. L. Turner II. 1992. Human population growth and global land-use/cover change. Annual Review of Ecology and Systematics. 23:39-61.

Opler, P. A. and R. L. Langston. 1968. A distributional analysis of the butterflies of Contra Costa County, California. Journal of the Lepidopterist's Society 22(2)89-107.

Parmesan, C. and N. Ryrholm, C. Stefanescu, J. K. Hill. C. D. Thomas, H. Descimon, B. Huntley, L. Kaila, J. Kullberg, T. Tammaru, W. J. Tennent, J. A. Thomas, and M. Warren. 1999. Poleward shifts in geographical ranges of butterfly species associated with regional warming. Nature 399:(6736)579-583.

Reynolds, R. T., J. M. Scott, and R. A. Nussbaum. 1980. A variable circular-plot method for estimating bird numbers. Condor 82:309-313.

SYSTAT 1992. SYSTAT: Statistics, Version 5.2 Edition. Evanston, IL, SYSTAT Inc.

Vitousek, P. M., H. A. Mooney, J. Lubchenco, and J. M. Melillo. 1997. Human domination of Earth's ecosystems. Science 277:494-499.

Vitousek, P. M., C. M. D'Antonio, L. L. Loope, and R. Westbrooks. 1996. Biological invasions as global change. American Scientist 84(5):468-478.

Wilcove, D. S., D. Rothstein, J. Dubow, A. Phillips, and E. Losos. 1998. Quantifying threats to imperiled species in the United States. BioScience 48(8):607 - 615.

Williams, J. E., J. E. Johnson, D. A. Hendrickson, S. Contreras-Balderas, J. D. Williams, M. Navarro-Mendoza, D. E. McAllister, and J. E. Deacon. 1989. Fishes of North America endangered, threatened, or of special concern: 1989. Fisheries 14(6):2 – 2.

Chapter 4

Rarity and Phylogeny in Birds

Thomas J. Webb[1], Melanie Kershaw[2] and Kevin J. Gaston[1]
[1]*Biodiversity & Macroecology Group, Department of Animal & Plant Sciences, University of Sheffield, Sheffield S10 2TN, UK.*
[2]*The Wildfowl & Wetlands Trust, Slimbridge, Gloucestershire GL2 7BT, UK.*

1. INTRODUCTION

It has been well documented that extinction risk in birds is not distributed randomly with regard to life history traits (e.g. Bennett & Owens 1997; Reed 1999). This, combined with the fact that closely related species will tend to share any heritable traits that make them extinction-prone (Gaston 1994; Gaston & Blackburn 1997a; McKinney 1997a), means that evolutionary branching will lead to taxonomic selectivity in extinctions, with some avian taxa far more likely than others to hold recently extinct or threatened species (Bennett & Owens 1997; Gaston & Blackburn 1997a; McKinney 1997a; Russell *et al.* 1998; Hughes 1999). Thus, the number of higher taxa (e.g. genera, families) lost for a given number of species extinctions is likely to be disproportionately high, especially if extinctions are more prevalent in monotypic genera (McKinney 1997b; Hughes 1999; Russell *et al.* 1998). Losing higher taxa means that any biological innovations unique to them will also be lost, leaving an impoverished, more homogeneous fauna (Myers 1997).

The focus of this chapter, though, is on phylogenetic patterns of rarity, rather than of extinction risk. This distinction raises two possibilities. First, rarity in birds may be phylogenetically clumped in much the same way as extinction risk, in which case anthropogenic impacts could be construed as merely the accentuation of natural processes. For instance, the probability of species with small population sizes going extinct is naturally elevated due to the greater vulnerability of small populations to, for example, environmental or demographic stochasticity (Raup 1991; Caughley 1994); human pressure might simply act to further increase this probability. This situation is intuitively appealing, as one might expect that the rarest species in any given taxonomic group will be those deemed to be at the highest risk of extinction. However, although it is tempting to view rarity and extinction risk as roughly synonymous, this is not necessarily the case, and the assumption that a species is at greater risk of extinction simply because it is rare is questionable (Mace & Kershaw 1997). Indeed, the dominant contributors to extinction risk in birds are extrinsic processes such as habitat destruction, rather than the sizes of their geographic ranges or populations *per se* (Collar *et al.* 1994). This raises the second possibility, namely that the taxonomic selectivity in extinction risk that we observe is *not* a simple consequence of phylogenetic patterns of rarity. Rather, it could arise due to the fact that the extrinsic factors driving extinctions tend to be concentrated in particular geographic areas, or in the same kinds of habitat and with similar outcomes (e.g. forms of fragmentation) in different regions, and thus impact upon related species in similar ways (Russell *et al.* 1998). In addition, because they are likely to have similar biologies, closely related species will tend to be susceptible to the same kinds of threats, to some extent regardless of their relative abundances. Thus, some bird species that are undoubtedly rare, such as many of those in the Costa Rica and Panama highlands and in several other Endemic Bird Areas (Stattersfield *et al.* 1998), are not currently considered to be threatened with extinction. Equally, previously common species, perhaps most famously the passenger pigeon *Ectopistes migratorius*, have been quickly driven to extinction by extrinsic factors (Bucher 1992); the dramatic declines in the abundances of many European birds provide further examples of how such dynamics might arise (Tucker & Heath 1994).

Such a decoupling of rarity and phylogeny would have serious implications for the future of avian diversity. The broad positive correlation between range size and time to extinction on both ecological and geological scales (see Gaston 1994, 1998; Rosenzweig 1995; McKinney 1997b) suggests that during times of background extinction rates, rare species will indeed be more likely than more widespread species to go extinct. If they are randomly spread across taxonomic lineages therefore, this natural extinction of rare species will cause a continual 'pruning' of the phylogenetic tree, removing terminal branches more or less at random. While this will evidently decrease species richness, evolutionary history can be preserved to a remarkable degree when extinctions are taxonomically random (Nee & May 1997); thus, most higher taxa (e.g. genera, tribes) will likely be left with at least one survivor to represent any biological or ecological innovations unique to these higher levels. With rarity independent of phylogeny, the current taxonomic selectivity in extinction risk, whose effect on a phylogenetic tree would (if carried to conclusion) be more akin to coppicing than pruning, could not simply be attributed to a higher risk of extinction in rarer taxa. The most obvious alternative explanation would be that, far from a simple exaggeration of natural processes, anthropogenic factors such as habitat destruction were instead *disrupting* these processes. The consequences of this for avian diversity would likely be severe and, with continuing human pressure, perhaps irreversible (Myers 1997).

Which, then, of these two possibilities (rarity is or is not randomly distributed with respect to phylogeny) seems more likely? Some studies have certainly hinted at a phylogenetic component to population size or range size. For example, there is some evidence that closely related species may have similar population densities within communities (Cotgreave & Harvey 1992); and in the Australian avifauna, there is more variation in the geographic range sizes of species between taxonomic orders than within them (Cotgreave & Pagel 1997). However, there are good reasons why we might often expect rarity to show no clear phylogenetic pattern (Cotgreave & Pagel 1997; and see section 3.2 below). Indeed, it is a general observation that geographic range sizes (and by inference population sizes) tend to be very variable within higher taxa, with marked differences in distributional extent a regular occurrence between species even within the same genus (Gaston 1998).

In this chapter, we further explore the relations between phylogeny and rarity in birds. First, we examine the distribution of rare taxa in one particularly well studied avian order, the Anseriformes. Then, by examining the mechanisms that may cause rarity in a natural system, we attempt to make a distinction between 'natural' and 'anthropogenic' rarity. This may provide insight into why extinction risk is not a simple function of abundance, by explaining how taxonomically selective, human-induced endangerment differs from natural patterns of rarity.

2. RARITY AND PHYLOGENY IN THE ANSERIFORMES

The avian order Anseriformes comprises the ducks, geese, swans and screamers. It is relatively speciose (163 extant species recognised by D.A. Callaghan, unpublished data), and is one of the best studied groups of organisms in the world (Madge & Burn 1989; Owen & Black 1990; Batt *et al.* 1992; del Hoyo *et al.* 1992; Todd 1996; D.A. Callaghan, unpublished data).

Several definitions of what constitutes rarity have been proposed in the literature, incorporating measures of, for example, global population size, local population density, geographic range size, habitat breadth, and temporal persistence (Rabinowitz 1981; Reveal 1981; Fiedler & Ahouse 1992; Gaston 1994, 1997). Some employ arbitrary divisions between 'rare' and 'common' species, and others treat rarity as a continuous trait. Throughout this chapter, we will employ continuous measures, based on abundance; we also only consider data that incorporate the entire global distribution of the species. For every extant species, as well as seven additional 'megasubspecies' (subspecies approaching species status), of the Anseriformes, the Wildfowl and Wetland Trust have estimated the global population size (taken here to be the maximum population size reported by

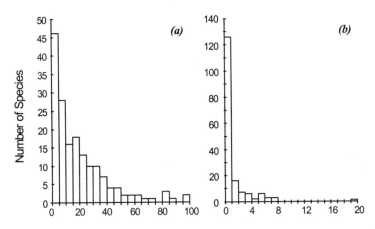

Figure 1. (a) Species-range size (number of WORLDMAP squares, each approximately 611 000 km^2, occupied during the breeding period; see text for details) and (b) species-population size (maximum number of individuals, x10^6) distributions for Anseriformes worldwide.

D.A. Callaghan, unpublished data) and the global breeding range size (the number of squares of a WORLDMAP grid occupied during the breeding period; each square represents an equal area (c.611 000 km^2) of 10° longitude; Williams 1996) from available published and unpublished data sources, census data and correspondence with appropriate experts. D.A. Callaghan (unpublished data) has subjectively evaluated the quality of the population size estimates as good, moderate or poor. Each category encompasses species with estimates ranging from under 1000 to several million individuals, providing no evidence for a systematic reduction in the quality of the estimates as population size increases; of course, a 'good' estimate of a species with a population of several million individuals may still be substantially less accurate than a 'poor' estimate of a species with between a few tens and a few thousands of individuals. Because of the relatively large scale of the WORLDMAP grid used, and because Anseriformes are so well studied, data quality is not likely to be an issue with the measures of geographic range size. Range size and population size are highly positively correlated across species in the wildfowl ($r=0.783$, $n=170$, $p<0.001$); this remains true when the phylogenetic relatedness of species is controlled for (Gaston & Blackburn 1996).

Abundance in the Anseriformes is extremely variable, some species being restricted to small islands whilst others breed on several continents (del Hoyo et al. 1992), a fact reflected in population sizes that span seven orders of magnitude. Nonetheless, it is immediately apparent that the majority of species are rare (figure 1): the smallest size class in both the species-range size and the species-population size distribution is also the modal one. This further highlights the fact that extinction risk is not the same thing as rarity, because although most anseriform species are rare, only 28 (16%) are currently considered to be threatened with extinction (D.A. Callaghan, unpublished data). Under logarithmic transformation, neither the species-range size nor the species-population size distribution differs significantly from a normal distribution (Kolmogorov-Smirnov tests: range size, $Z=1.1148$, $p=0.1665$; population size, $Z=1.2130$, $p=0.1054$). These patterns in the shapes of species-range size and species-population size distributions are fairly general, and have been observed previously for Anseriformes, as well as for other taxonomic groups of birds and for other organisms (Anderson 1984a, b; Pagel et al. 1991; Gaston 1994, 1996, 1998; Blackburn & Gaston 1996; Gaston & Blackburn 1996; Blackburn et al. 1998; Gaston & Chown 1999).

Groups of closely related species will tend to share many traits as a result of their common evolutionary history, regardless of their present ecological circumstances. Such traits, which are present in an organism because they were present in its ancestor, rather than due to any ongoing adaptive value, are said to be phylogenetically constrained (McKitrick 1993; Johnson et al. 1999). If, therefore, abundance were heritable and phylogenetically constrained, then closely related species should have similar abundances through common descent, and we would expect much of the variation in abundance across species to be explained at higher taxonomic levels, such as the tribe. On the other hand, if abundance really does behave differently from many other biological traits, then phylogenetic inertia will play only a small role in determining the abundances of species. In this case, we would expect to see high levels of variability in the abundances of more closely related species, for example among congeneric species.

For the Anseriformes, the amount of variation in abundance that was explained at each of the levels of subfamily, tribe, genus and species (following the classification used by D.A. Callaghan (unpublished data), which is based on that of Sibley & Monroe 1990) was estimated using a nested ANOVA. Data for body size and clutch size were also considered, so that patterns in abundance could be compared with those in other biological traits. Because of the unbalanced nature of the ANOVA, F and p values are not reported. However, it is clear from figure 2 that the link between rarity and phylogeny in wildfowl, as in other groups of animals (Blackburn *et al.* 1998; Gaston & Blackburn 1997b; Gaston 1998), is extremely tenuous; most variance is only explained at the level of the species (figure 2a, b). This is true whether population size or geographic range size is considered, and contrasts strongly with the patterns observed in the other traits analysed, where much of the variance is explained at the level of the genus, tribe or even subfamily (figure 2c, d). In other words, whilst generalisations such as "swans are large-bodied" or "pochards have large clutches" could be justified, no such sweeping statements can be made about abundance. If, then, as seems likely, there is very little phylogenetic inertia in abundance, does this mean that rarity is entirely unrelated to phylogeny? Do rare species

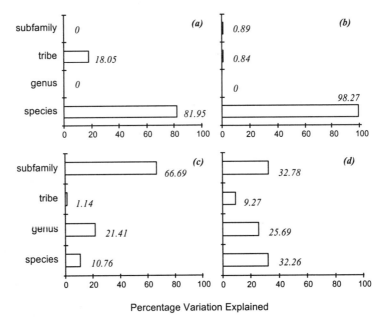

Figure 2. The percentage of variation in *(a)* breeding geographic range size, *(b)* maximum population size, *(c)* body size (mean female body mass, g) and *(d)* average clutch size explained at different taxonomic levels in the Anseriformes worldwide. Variance components are estimated using a nested ANOVA, and negative variance estimates are replaced by zero (following Rao & Heckler 1997).

owe their circumstance more to chance than to evolutionary history, thus implying that the taxonomic selectivity in extinction risk has been artificially imposed? In order to address this question, it is necessary to look at the processes that are operating to keep the pool of globally rare species so full.

3. ABUNDANCE AND EVOLUTIONARY HISTORY

The observed species-range size distribution (and by inference, the species-population size distribution) on a global scale must be a product of patterns of speciation and extinction, and of the evolution of geographic range sizes of individual species between their birth and their demise (Price *et al.* 1997; Gaston 1998; Gaston & Chown 1999). There are several ways in which these mechanisms might operate to produce the patterns observed in the Anseriformes, not all of which necessarily require that the determination of abundance has no phylogenetic component. For instance, a species might inherit its range size from its ancestor at speciation, leading to the clustering of rare species within certain clades; this pattern may then be obscured by subsequent transformation of the geographic ranges of species within the 'rare' clade. Such transformations would presumably not follow a strict pattern, because if the ranges of all species within a clade evolved in an identical manner, at any given point in time we would expect to still see some similarity between them. Alternatively, speciation may divide geographic ranges at random: with or without post-speciation transformation of range sizes, this lack of heritability of range sizes would translate into the present day pattern of both rare and common species being ubiquitous across the phylogeny. Here, we examine how these natural processes of speciation and range size transformation allow high levels of diversity to be maintained, and contrast this with the destructively selective impact that humans are having upon avian diversity.

3.1 Speciation

Speciation generates new species, and hence additional geographic ranges (Gaston 1998; Gaston & Chown 1999). Here, we assume that speciation is predominantly allopatric (Anderson & Evensen 1978; Mayr 1982; Peterson *et al.* 1999), whilst acknowledging that other modes may be important, at least in some groups of organisms (e.g. Tauber & Tauber 1989; Ripley & Beehler 1990; Schliewen et al. 1994; Rosenzweig 1995; Dieckmann & Doebeli 1999). In such a situation, the immediate effect that these new ranges will have on the pool of rare species as reflected in the shape of the species-range size distribution will depend upon the size of the ancestral geographic range, and the way in which it is divided. One possibility that has been raised, and widely cited, is that new species inherit much of their geographic range size from their immediate ancestor (Jablonski 1987). This would result in a strong relationship between the geographic range sizes of pairs of sister species. To investigate this possibility in the Anseriformes, 46 pairs of sister taxa were identified from the phylogenies of Livezey (1986, 1991, 1995a,b,c, 1996a,b, 1997). The order of species within each of the pairs was randomised, and Spearman rank correlations were determined across the pairs between the two geographic range sizes. The values of r_s obtained for 100 randomisations were approximately normally distributed with a mean of 0.180±0.002 (S.E.); a typical result is shown in figure 3a. None of the correlation coefficients obtained was significant at the 5% level. These correlations were contrasted with those obtained from an identical analysis for a life history trait (body size). The correlation coefficients obtained in this analysis were again approximately normally distributed, but this time around a mean of 0.940±0.001. Each one of the 100 values of r_s obtained for the body size analysis was highly significant (P<0.001); a typical result is shown in figure 3b. These results suggest once again that abundance does not behave like other biological traits, and that whereas sister taxa tend to have very similar body sizes (figure 3b), there is no significant relationship between the geographic range sizes of very closely related species (figure 3a). This supports the conclusions of Anderson and Evensen (1978), who found no evidence that the ratio of range sizes within 53 sister-taxa pairs of North American mammals, birds and amphibians was anything other than random. (N.B. as expected, given the strong correlation between geographic range size and population size in the Anseriformes (see above), our conclusions hold if population size is used as a measure of abundance: mean r_s=0.186±0.002, n=100; no significant correlations).

Using the results of this analysis to test for the inheritance of range size from ancestral to daughter species assumes that the geographic range sizes of species today reliably reflects the range sizes that they inherited at speciation. This is clearly unrealistic, precluding as it does the possibility that range sizes change post-speciation (although the significance of such change and its influence on the extent to which the present-day distributions of species can be used to reconstruct past patterns of speciation has been much debated; Lynch 1989; Brooks & McLennan 1991). Whilst our analyses confirm that there is little correlation between the range sizes of sister taxa today, they do not necessarily reflect a lack of heritability of range size at speciation. However, it does seem unlikely that sister taxa will share similar range sizes even immediately following speciation, given that many

4. Rarity and Phylogeny in Birds

geographic models of allopatric speciation imply that ancestral range sizes will not be divided equally at speciation (Gaston 1998; Gaston & Chown 1999). The most obvious example is the highly asymmetrical split in the ancestral range, leading to a daughter species with a small range size, which will result from speciation in small peripheral isolates (Glazier 1987; Price *et al.* 1997; Gaston & Chown 1999). Vicariance, too, will tend to produce two daughter species with different range sizes, although in this case it is at least feasible that range sizes will be inherited, as any degree of asymmetry in the division of ranges is possible. The likelihood of an apparent heritability of

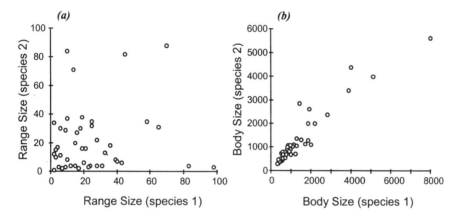

Figure 3. Typical examples of the comparisons of *(a)* geographic range sizes (number of WORLDMAP squares, each approximately 611 000 km^2, occupied during the breeding period; see text for details) and *(b)* body sizes (female body mass, g) in 46 pairs of anseriform sister taxa. There is no significant correlation of range sizes within sister taxa (Spearman rank correlations, mean $r_s=0.180\pm0.002$ (S.E.), n=100; this example $r_s=0.202$, p=0.179). In contrast, sister taxa tend to have very similar body sizes (mean $r_s=0.940\pm0.001$; this example $r_s=0.937$, p<0.001).

range sizes will depend on the range size of the ancestral species. For example, if a species with a large range speciates, then the product could span from two daughter species with very different range sizes (one large and one small) to two daughter species with identical range sizes (the result of a perfectly symmetrical range division). The most likely result, however, is two daughter species with different sized distributions: there are many ways to split a range unequally, but only an exact 50:50 split will create two ranges of identical size (Gaston 1998; Gaston & Chown 1999). It is only when the ancestral range size is small that the newly created sister taxa will be sure to have similar range sizes, as the product of any range division can only be two small ranges. This effect will be exacerbated by the relatively crude way that the range sizes of anseriform species have been measured in the data employed here (where a range size can only take an integer value). As there are n-1 ways of dividing an ancestral range of size n into two daughter ranges, both taking positive integer values, there are fewer ways to divide a small range than a large range; and because all small ranges will split to form two small ranges, whilst *some* large ranges will split to form two moderately large ranges, this may lead to a positive correlation between the range sizes of daughter taxa immediately following speciation even if ancestral range sizes are split entirely at random. The large ranges that split in a highly asymmetrical fashion will be the noise around this relationship. Even if a marginally significant correlation between the range sizes of sister taxa is found, then, it should be interpreted with caution; and it wouldn't necessarily prevent us from concluding that in most cases, irrespective of the relative frequency of peripheral isolation and vicariance models (a point of some contention; e.g. Bush 1975; Barton & Charlesworth 1984; Mayr 1988; Lynch 1989; Brooks & McLennan 1993; Frey 1993; Ripley & Beehler 1993; Chesser & Zink 1994; Taylor & Gotelli 1994; Wagner & Erwin 1995; Dimmick et al. 1999), we would expect low levels of heritability in geographic range size, as suggested by the anseriform data.

One further aspect of speciation that may influence patterns of abundance across a phylogeny is the form of the relationship between likelihood of speciation and geographic range size. It seems probable that the likelihood of speciation is not an unbiased function of range size, although the precise pattern of bias has been much discussed, with conflicting views being expressed (Darwin 1859; Mayr 1963, 1988; Rosenzweig 1978, 1995; Stanley 1979; Chown 1997; Gaston 1998; Gaston & Chown 1999; Maurer 1999). If the effects of asymmetrical divisions in geographic ranges are more pronounced when the ancestral range size is large, it follows that they will be of less importance to the species-range size distribution if rare species are more likely to speciate than common ones. With speciation a regular occurrence in rare species, rarity would likewise be a regular outcome of speciation (Chown 1997), and we might expect to see more congruence in the range sizes of sister taxa than in fact appears to be the case. If the intrinsic properties of species, which tend to be phylogenetically constrained, directly influence the probability of speciation (Chown 1997), then we might also expect to see taxonomic selectivity in speciation. It should follow from this that, as speciation tends to produce rare species, rarity will also be more prevalent in certain taxa.

The fact that such patterns have not emerged would seem to support the view that speciation is more likely in widespread and abundant taxa (Rosenzweig 1978, 1995). However, this need not be the case. For example, the bias towards speciation in taxa possessing certain life history traits may not apply to all modes of speciation (Barraclough *et al.* 1999; Dimmick *et al.*

1999). Two further possibilities are that even if rare species are more likely to undergo allopatric speciation, because of their shorter longevity (see below) they may on average contribute fewer daughters (Gaston & Chown 1999), and these daughters (which logically must be rarer still) may be unlikely to persist (Glazier 1987). Therefore, even though the instantaneous rate of speciation may be lower in common taxa, the fact that they tend to produce more daughter taxa and daughters with larger range sizes may mean that most of the species which exist for us to study have in fact resulted from such speciation events, thus inflating the importance of range division asymmetries. Alternatively, patterns that exist at speciation may be obscured (or even obliterated) by the subsequent transformation of range sizes (Glazier 1987; Chesser & Zink 1994; Barraclough et al. 1998). Such changes in range size may also help to reconcile studies of contemporary taxa, which typically show very little heritability of range size (Gaston 1998), with the results of paleontological studies which have found such heritability (e.g. Jablonski 1987). Contemporary taxa of the same age may have different abundances because their range sizes are changing at different rates, and the rarer species today may have historically had a similar range size to that of its common relative, or may achieve such a range size at some future point. If this was the case, then at any one time there would be little relationship between the range sizes of closely related species, but strong interspecific patterns in range sizes may become apparent when they are summed over evolutionary time (Gaston 1998). Such summation is effectively what happens in paleontological studies (because of the relatively crude temporal resolution of the fossil record, and also if range sizes used in studies are summed across strata), which may therefore partially control for the effects of range size transformations.

3.2 Range Transformations

We have seen that post-speciation transformations in geographic range sizes will lessen the effects of the mode of speciation on the species-range size distribution, and can help to explain some of the observed patterns in abundance. It seems reasonable to expect that, particularly in species with high dispersal abilities such as birds (Chesser & Zink 1994), range sizes will change as species respond (or fail to respond) to changes in ecological and environmental circumstances (Price et al. 1997; Barraclough et al. 1999). This may well occur over time periods that would not permit drastic changes in morphology or life history to occur (Blackburn et al. 1996; Kunin 1997). Indeed, to some extent range sizes must change post-speciation: all models of allopatric speciation necessarily cause a reduction in range sizes (Glazier 1987), and so if they were fixed at speciation, then we would expect to see a progressive decline in the average range sizes of species through evolutionary time, as they were continually subdivided (Gaston 1998).

Several models have been proposed of the form that transformations in geographic range sizes take (see Gaston & Blackburn 1997a; Gaston & Kunin 1997; Gaston 1998; Gaston & Chown 1999). For example, a species could quickly achieve its destined abundance and maintain it until it crashes to extinction (a 'stasis' model; Jablonski 1987); in other words, its range size is essentially fixed for the duration of its existence. Alternatively, newly created species may tend to be rare, and to increase gradually in abundance over time until eventually crashing to extinction (an 'age and area' model; Willis 1922); or conversely, they may rapidly increase to their maximum abundance and then decrease gradually until extinction. Another possibility is that species gradually increase to a peak in abundance at some intermediate age, before gradually decreasing towards extinction. If species within a higher taxon tend to conform to one of these models, then rarity would be expected (or 'natural') at some point during a species' lifespan. A species with a range much lower than that predicted for its age by the appropriate model may be rare for another reason, perhaps anthropogenic in origin; this might be termed 'artificial' rarity. If on the other hand there is no consistent pattern, with abundance changing rapidly and unpredictably between speciation and extinction (a reasonable 'null' model), then this method will be uninformative as to the 'naturalness' or otherwise of a species' predicament.

To rigorously test which of these patterns of transformations is operating in any particular case, we would need detailed information on the abundance of each species from the moment it differentiated until its extinction (Gaston & Kunin 1997). Of course, due to the imperfections of the fossil record, particularly as applied to the study of abundance (McKinney 1997b), this kind of data will be unattainable in almost every case. To the extent that broad interspecific patterns will reflect an intraspecific relationship (Gaston 1998), however, reliable molecular phylogenies may provide us with an insight into the form that past range size transformations might have taken. If the genetic divergence between extant species in a phylogeny is assumed to have accumulated in a clocklike manner, then nodes within a phylogeny can be dated and taxon age can be estimated. Plotting the present distributions of species against taxon age will then give an idea of the interaction between evolutionary age and geographic range size.

There are problems with this approach, notably the assumption of a molecular clock, which is a contentious issue (Klicka & Zink 1997); clocklike divergence may be affected, for example, by differential generation

4. Rarity and Phylogeny in Birds

times among taxa (Mooers & Harvey 1994). Such problems can be minimised by only considering groups of closely related species; as well as the fact that such groups are likely to be similar with regards to the biological factors that may bias molecular evolution, differences between closely related species tend to occur in a region of the genome which probably does evolve in a clocklike manner (Klicka & Zink 1997). Restricting study to narrowly delimited groups of species also increases the chance that the phylogeny used will be accurate and complete. This is an assumption of the analysis, because in using the date of the split between two neighbouring species on the phylogeny as a measure of taxon age, we are assuming that there is no 'missing' species (either an extant species not included in the phylogeny, or an unknown extinction) that should be placed between them. Such species have the potential seriously to affect estimates of taxon age. The approach adopted here, then, is to analyse separately groups of closely related species for which a phylogeny is available: the albatrosses (family Diomedeidae: Nunn *et al.* 1996), gannets and boobies (family Sulidae: Friesen & Anderson 1997), and warblers of the genera *Acrocephalus* and *Hippolais* (family Sylviidae: Helbig & Seibold 1999).

Figure 4 shows plots of taxon age versus breeding range size for these groups of birds. In all three cases, it could be argued that geographic range size is following one of the models outlined above, rather than being randomly distributed with respect to taxon age. For the two seabird families (figure 4a, b), it appears that a rapid expansion of geographic range size post speciation is followed by a gradual decline thereafter; whereas in the warblers (figure 4c), there appears to be some suggestion of a more gradual expansion followed by a gradual contraction in range sizes. It is unclear why we should observe this gradual decline in the range sizes of common species. Perhaps widespread taxa might be more susceptible to range fragmentation, making them vulnerable to displacement by younger expanding taxa (Wilson 1961). Alternatively, it could be that as a large range fragments, newly isolated populations at the periphery diverge into separate species. Several occurrences of this cladogenesis with ancestral persistence (Gaston & Chown 1999) would cause a gradual decline in the range size of the ancestral species as they passed from middle age to old age, leading to the oldest taxa only persisting as relicts (Wilson 1961). Whatever the precise mechanism, however, it appears that rarity is a frequent and natural occurrence in old taxa, as well as in young taxa (thus supporting the notion that speciation regularly generates rare species; Chown 1997). This may explain to some extent why, when examining a particular avian order, patterns of rarity are so unclear: higher taxonomic levels such as tribes will contain species of various evolutionary ages.

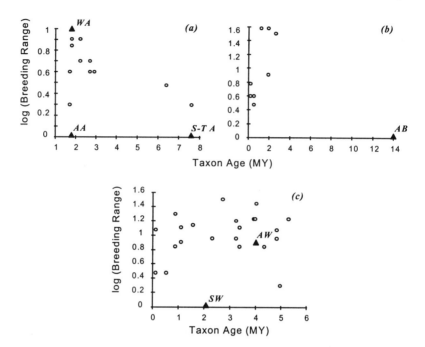

Figure 4. Breeding geographic range size as a function of taxon age. *(a)* Albatrosses (family Diomedeidae); *(b)* gannets and boobies (family Sulidae; after Friesen & Anderson 1997); *(c)* warblers (family Sylviidae) of the genera *Acrocephalus* and *Hippolais*. Solid triangles indicate species listed as threatened with extinction in Collar *et al.* (1994): *WA*, wandering albatross *Diomedea exulans*; *AA*, Amsterdam albatross *D. amsterdamensis*; *S-T A*, short-tailed albatross *D. albatrus*; *AB*, Abbott's booby *Papasula abbotti*; *SW*, Seychelles warbler *Acrocephalus sechellensis*; *AW*, Aquatic warbler *A. paludicola*. Estimates of taxon age are given for the gannets and boobies in Friesen & Anderson (1997), and for the albatrosses were calculated from the pairwise genetic divergences and rate of divergence given in Nunn *et al.* (1996). For the warblers, taxon ages are estimated from the data given in Helbig & Seibold (1999) under the assumption of a mtDNA substitution rate of 2% per million years, which applies to a diverse array of avian orders (see Klicka & Zink (1997) and references therein). Breeding range sizes are the number of WORLDMAP squares (each with an area of c. 611 000 km^2) occupied during the breeding period, and are calculated from distribution maps given in Harrison (1985) and Cramp (1992).

If changes in geographic range size do not follow an idiosyncratic pattern through time, how do we account for the fact that sister taxa (which are by definition the same age) have very different range sizes (figure 3)? Gaston (1998) explains how the patterns that we observe between sister taxa could be generated if both ranges expanded post-speciation, but then arrived at different maximum extents. Equally, the range sizes could be expanding at different rates towards the same, or towards different, maximum extents.

4. Rarity and Phylogeny in Birds

Indeed, the fact that any pattern emerges from plots of taxon age versus geographic range size is encouraging, because of the masking effects that quantitative differences in the rates or extents of range expansion or contraction will have even on qualitatively identical patterns of range size transformation; and enables us to begin to make the distinction between 'natural' and 'artificial' rarity.

3.3 'Natural' versus 'Artificial' Rarity

Figure 4 suggests that species which are young or old when compared to their close relatives might be expected to be rare, due to their natural position on a taxon age-range size curve; and that this might not necessarily put them at any great risk of extinction. Indeed, there are examples in figure 4 of young and old species with small breeding range sizes, which are nonetheless not considered to be threatened with extinction. This might be termed 'natural rarity'. Possible distinctions between young and old species can be examined by considering those that *are* threatened, and by comparing the criteria by which they have been listed as such by Collar *et al.* (1994). Young species are expected to have never been abundant: their fate will either be to embark upon a period of range expansion, or (perhaps more likely) to decline swiftly to extinction. In any case, they are likely to be considered threatened purely because of their restricted distribution, as is indeed the case in the Amsterdam albatross *Diomedea amsterdamensis* (figure 4a). Old species, on the other hand, are expected to be declining in abundance from previously higher levels, perhaps because they are poorly adapted to current conditions. This might make them more susceptible to the extrinsic factors that lead to increased extinction risk, which may in turn accelerate their decline. In both of the old species which are listed as threatened, the short-tailed albatross *D. albatrus* (figure 4a) and Abbott's booby *Papasula abbotti* (figure 4b), the primary reason given for inclusion on the list of threatened species is a marked decline from previous levels of abundance.

Species of intermediate age but with smaller range sizes than expected may be more likely to be at risk of extinction than rare species that are especially old or young. They fall into two groups: some may have managed to persist despite never having increased from a low abundance at speciation. This seems likely to be the case with the Seychelles warbler *Acrocephalus sechellensis* (figure 4c). Alternatively, they may be previously widespread species that have suffered a reduction in abundance due to extrinsic factors, as appears to be the case for the aquatic warbler *Acrocephalus paludicola* (figure 4c), which has suffered from destruction of its habitat (Collar *et al.* 1994). The wandering albatross *Diomedea exulans* (figure 4a) may also fall into this category; it has suffered significant population declines (c.50% in 20-30 years) primarily due to the drowning of birds on tuna longlines (Collar *et al.* 1994), although its breeding range size, as scored using the crude approach employed here, is still relatively high because it breeds on several small but widely separated islands (Harrison 1985). Examination of more data sets will show whether rare species of intermediate age will always be considered threatened. The phylogeny used for the warblers, for example, has several missing taxa, including threatened species; one might predict that at least some of these will fall in the same region of the taxon age-range size plot as the Seychelles warbler.

If, as we have seen above, taxa often tend to inherit small range sizes at speciation and, after a period of expansion, then gradually decline in range size again as they age, the extinction of young species (which become extinct before their ranges are able significantly to expand) and old species will be a frequent natural occurrence. Thus natural extinctions will have little effect on overall levels of biodiversity, as such extinctions will primarily occur in species with small geographic range sizes (see Gaston 1994, 1998; Rosenzweig 1995; McKinney 1997b), and small geographic range sizes will occur in species from all taxonomic groups.

If human activity was merely accentuating this natural pattern, perhaps by suppressing the expansion of the ranges of young species, maintaining rarity into intermediate ages, and hastening the decline of old species, then there would be no reason to expect taxonomic selectivity in extinction risk. The fact that such a pattern does exist, therefore, suggests that humans are not simply targeting naturally rare species. Indeed, a feature of mass extinction events, such as the one being precipitated by current human activity (Ehrlich & Ehrlich 1981), is that the correlation between geographic range size and likelihood of extinction breaks down (Jablonski 1986; Norris 1991); and we have only to look at the examples littering human history of formerly widespread species that have rapidly become extinct or threatened with a high likelihood of extinction in the near future (e.g. Vermeij 1993; Lawton 1995; Rosenzweig 1995) to appreciate that having a large geographic range size does not always protect a species from extinction. For example, by the 1930s hunting pressure had reduced the trumpeter swan *Cygnus buccinator*, which had been widespread and abundant in northern North America until the mid nineteenth century, to the brink of extinction (Madge & Burn 1989; Owen & Black 1990; Mitchell 1994), although it has now recovered thanks to a conservation programme. This situation is by no means unique to birds. For instance, intensive fishing has reduced some 70% of global marine fish stocks to the extent that they are now in need of urgent management (Anon 1997), and previously common species such as the barndoor skate *Raja laevis* are now close to extinction (Casey & Myers 1998).

It is possible that species with a history of rarity will evolve adaptations to life at low abundance (Kunin 1997; Orians 1997). If this is the case, then the picture is still bleaker for recent, artificial rarities: lacking such adaptations, they will be at an increased risk of extinction when compared to naturally rare species with a similar abundance (Lawton 1995; Gaston & Kunin 1997). This may partially explain why Manne *et al.* (1999) found that, for a given range size, New World passerine birds that live on islands are less likely than their continental relatives to be threatened with extinction: the island birds have adapted to life with a restricted range, whereas the continental birds, whose ranges have been recently reduced due to fragmentation of their forest habitats, have no such adaptations. As outlined in the introduction, such artificial rarity affects a non-random subset of taxa (Bennett & Owens 1997; McKinney 1997b; Russell *et al.* 1998), which means that the resultant extinctions and endangerments will cause a disproportionately high disappearance of higher taxa (McKinney 1997b; Russell *et al.* 1998), and consequently exaggerate the loss of diversity.

4. CONCLUSIONS

In birds, as in other taxa, rarity is a widespread and natural phenomenon. Our analyses on the Anseriformes have shown that, unlike the biological traits of individuals, which tend to be highly conserved between closely related species, rarity is not strongly constrained by phylogeny. This fact could arise because abundance is not partitioned equally at speciation, which is a likely outcome of certain modes of speciation. Alternatively, changes in abundance might occur subsequent to speciation. We have shown that within avian families, abundance does not appear to be a random function of taxon age. Rather, post-speciation transformations in range size may possibly follow some predictable trajectory, with young and old taxa tending to have lower abundances than species of intermediate age.

These results highlight the fact that extinction risk and rarity are not synonymous. Species can be rare without being considered to be unduly threatened with extinction, and this might tend to apply to 'naturally' rare young and old taxa. Human disturbance, on the other hand, has brought many previously abundant species to, or in some cases over, the brink of extinction (Rosenzweig 1995). Such 'artificial' rarity may occur in species of any age, by affecting *(a)* which rare species are able to expand their ranges, *(b)* which common species will suffer a decline in abundance, and *(c)* which naturally declining species are hastened on their path to extinction.

This human factor tends to affect closely related species in similar ways (Russell *et al*. 1998), and so we observe phylogenetic patterns of extinction risk in birds (e.g. Bennett & Owens 1997; Hughes 1999; Reed 1999). Given that both population size and geographic range show no such phylogenetic pattern in birds, it seems likely that other characteristics which *are* phylogenetically heritable (such as body size or reproductive effort; Bennett & Owens 1997) will contribute more to the non-random allocation of extinction risk. It is interesting to note that, like extinction risk, invasiveness (as measured by introduction success) in birds also shows taxonomic selectivity (Lockwood 1999; McKinney & Lockwood 1999). Here too, ecological specialisation (which will be influenced by factors that are heritable, such as physiology and diet; McKinney 1998) seems to play a greater role in influencing a species' chance of being successfully introduced (with a generalist more likely to be a successful invader than a specialist) than either its population size within its native geographic range, or the size of that range (Brooks, this volume).

We therefore arrive at a distinction between a 'natural' situation and the situation which prevails today. In the former, rarity occurs at random across the avian phylogeny. Although many of these rare species may go extinct by chance, biodiversity will be maintained because the underlying structure of the phylogenetic tree is maintained (Nee & May 1997); the addition of more species through speciation may even lead to an increase in diversity. In the current situation, however, human actions may have contributed to the taxonomically non-random patterns of extinctions and extinction risk, with the result that not only are certain species at a higher risk of disappearing than others, but so too are the higher taxa to which they belong (Bennett & Owens 1997; Russell *et al*. 1998). To the extent that related species will tend to be exposed to similar environments (Russell *et al*. 1998), the majority of species within a clade may suffer comparable consequences from any natural disturbance, so causing taxonomic clumping of extinctions. The disproportionate loss of biodiversity resulting from human activity may therefore not be completely unprecedented; indeed, it has been documented that natural patterns of extinction tend to break down during periods of mass extinction (Jablonski 1986; Norris 1991). The frequency and intensity of human disturbance, however, may inhibit the recovery of biodiversity, which has previously relied on the origination of new species by speciation (a relatively slow process; McKinney 1998). This effect will only be exacerbated by the enormous scale of the worldwide spread of species today facilitated by humans (Vitousek *et al*. 1997). The primary foci of biological invasions tend to be human-altered ecosystems (Vitousek *et al*. 1997), and so the rapid establishment of exotic species in such areas, even when they themselves are not directly responsible for the extinction of local endemics, may remove or lessen the conditions necessary for speciation. The fact that avian introductions are biased towards certain taxonomic groups (Lockwood 1999), and particularly towards ecological generalists (Brooks, this volume), means that the loss of distinctive taxonomic groups across the world will tend to benefit ecologically similar species everywhere (McKinney & Lockwood 1999). This constitutes an extensive and sustained disruption of natural patterns of rarity and diversity and will accelerate the process of biotic homogenisation (Lockwood, 1999; McKinney & Lockwood 1999), certainly resulting in an impoverished avifauna. As Lawton (1995, p. 159) states, "[we] may not wish to destroy gorillas and bowerbirds and encourage starlings and cockroaches. But we seem powerless to do anything about it." The worry with the present wave of extinctions is that continuing human-imposed selection for such a 'pest and weed' ecology (Myers 1997) may mean that distinctive ways of life, among the birds and among other taxa, are lost forever.

ACKNOWLEDGEMENTS

We are grateful to two anonymous referees for comments, to Steven Chown, Nigel Collar and Ana Rodrigues for discussion and comments, and to Des Callaghan and Rebecca Woodward for compiling and inputting the Anseriformes data. T.J.W. is supported by a WhiteRose studentship. K.J.G. is a Royal Society University Research Fellow.

REFERENCES

Anderson, S. 1984a. Geographic ranges of North American terrestrial birds. American Museum Novitates 2785:1-17.
Anderson, S. 1984b. Areography of North American fishes, amphibians and reptiles. American Museum Novitates 2802:1-16.
Anderson, S. and Evensen, M. K. 1978. Randomness in allopatric speciation. Systematic Zoology 27:421-430.
Anon. 1997. Worldfish Report. London: Agra Europe Ltd.
Barraclough, T. G., Hogan, J. E. and Vogler, A. P. 1999. Testing whether ecological factors promote cladogenesis in a group of tiger beetles (Coleoptera: Cicindelidae). Proceedings of the Royal Society of London Series B-Biological Sciences 266:1061-1067.
Barraclough, T. G., Vogler, A. P. and Harvey, P. H. 1998. Revealing the factors that promote speciation. Philosophical Transactions of the Royal Society of London Series B-Biological Sciences 353:241-249.
Barton, N. H. and Charlesworth, B. 1984. Genetic revolutions, founder effects, and speciation. Annual Review of Ecology and Systematics 15:133-164.

Batt, B. D. J., Afton, A. D., Anderson, M. G., Ankney, C. D., Johnson, D. H., Kadlec, J. A. and Krapu, G. L. (ed.) 1992. Ecology and Management of Breeding Waterfowl. University of Minnesota Press, Minneapolis.

Bennett, P. M. and Owens, I. P. F. 1997. Variation in extinction risk among birds: chance or evolutionary predisposition? Proceedings of the Royal Society of London Series B- Biological Sciences 264:401-408.

Blackburn, T. M. and Gaston, K. J. 1996. Spatial patterns in the geographic range sizes of bird species in the New World. Philosophical Transactions of the Royal Society of London Series B- Biological Sciences 351:897-912.

Blackburn, T. M., Gaston, K. J. and Lawton, J. H. 1998. Patterns in the geographic ranges of woodpeckers (Aves: Picidae). Ibis 140:626-638.

Blackburn, T. M., Lawton, J. H. and Gregory, R. D. 1996. Relationships between abundances and life histories of British birds. Journal of Animal Ecology 65:52-62.

Brooks, D. R. and McLennan, D. A. 1991. Phylogeny, Ecology, and Behaviour: A Research Program in Comparative Biology. University of Chicago Press, Chicago.

Brooks, D. R. and McLennan, D. A. 1993. Comparative study of adaptive radiations with an example using parasitic flatworms (Platyhelminthes: Cercomeria). American Naturalist 142:755-778.

Bucher, E. H. 1992. The causes of extinction of the passenger pigeon. Current Ornithology 9:1-36.

Bush, G. L. 1975. Modes of animal speciation. Annual Review of Ecology and Systematics 6:334-364.

Casey, J. M. and Myers, R. A. 1998. Near extinction of a large, widely distributed fish. Science 281:690-692.

Caughley, G. 1994. Directions in conservation biology. Journal of Animal Ecology 63:215-244.

Chesser, R. T. and Zink, R. M. 1994. Modes of speciation in birds: a test of Lynch's method. Evolution 48:490-497.

Chown, S. L. 1997. Speciation and rarity: separating cause from consequence. In: The Biology of Rarity - Causes and Consequences of Rare-Common Differences (W. E. Kunin and K. J. Gaston, eds.), Chapman and Hall, London, Pp: 91-109.

Collar, N. J., Crosby, M. J. and Stattersfield, A. J. 1994. Birds to Watch 2: The World List of Threatened Birds. BirdLife International, Cambridge.

Cotgreave, P. and Harvey, P. H. 1992. Relationships between body size, abundance and phylogeny in bird communities. Functional Ecology 6:248-256.

Cotgreave, P. and Pagel, M. 1997. Predicting and understanding rarity: the comparative approach. In: The Biology of Rarity - Causes and Consequences of Rare-Common Differences (W. E. Kunin and K. J. Gaston, eds.), Chapman & Hall, London, Pp: 236-261.

Cramp, S. (ed.) 1992. The Birds of the Western Palearctic. Oxford University Press, Oxford.

Darwin, C. 1859. On the Origin of Species by Means of Natural Selection, or the Preservation of Favoured Races in the Struggle for Life. John Murray, London.

del Hoyo, J., Elliott, A. and J. Sargatal (ed.) 1992. Handbook of the Birds of the World. Lynx Edicions, Barcelona.

Dieckmann, U. and M. Doebeli 1999. On the origin of species by sympatric speciation. Nature 400:354-357.

Dimmick, W. W., M.J. Ghedotti, M.J. Grose, A. M. Maglia, D.J. Meinhardt, and D. S. Pennock 1999. The importance of systematic biology in defining units of conservation. Conservation Biology 13:653-660.

Ehrlich, P. R. and A.H. Ehrlich 1981. Extinction. The Causes and Consequences of the Disappearance of Species. Random House, New York.

4. Rarity and Phylogeny in Birds

Fiedler, P. L. and J.J. Ahouse 1992. Hierarchies of cause: toward an understanding of rarity in vascular plant species. In: Conservation Biology: the Theory and Practice of Nature Conservation, Preservation and Management (P. L. Fiedler and S. K. Jain, eds.), Chapman and Hall, London, Pp: 23-47.

Frey, J. K. 1993. Modes of peripheral isolate formation and speciation. Systematic Biology 42:373-381.

Friesen, V. L. and D.J. Anderson 1997. Phylogeny and evolution of the Sulidae (Aves: Pelecaniformes): a test of alternative modes of speciation. Molecular Phylogenetics and Evolution 7:252-260.

Gaston, K. J. 1994. Rarity. Chapman & Hall, London.

Gaston, K. J. 1996. Species-range-size distributions: patterns, mechanisms and implications. Trends in Ecology and Evolution 11:197-201.

Gaston, K. J. 1997. What is rarity? In: The Biology of Rarity - Causes and Consequences of Rare-Common Differences (W. E. Kunin and K. J. Gastont eds.), Chapman & Hall, Pp: 30-47.

Gaston, K. J. 1998. Species-range size distributions: products of speciation, extinction and transformation. Philosophical Transactions of the Royal Society of London Series B-Biological Sciences 353:219-230.

Gaston, K. J. and Blackburn, T. M. 1996. Global scale macroecology: interactions between population size, geographic range size and body size in the Anseriformes. Journal of Animal Ecology 65:701-714.

Gaston, K. J. and Blackburn, T. M. 1997a. Evolutionary age and risk of extinction in the global avifauna. Evolutionary Ecology 11:557-565.

Gaston, K. J. and Blackburn, T. M. 1997b Age, area and avian diversification. Biological Journal of the Linnean Society 62:239-253.

Gaston, K. J. and Chown, S. L. 1999 Geographic range size and speciation. In: Evolution of Biological Diversity (A. E. Magurran and R. M. May, eds.), Oxford University Press, Oxford, Pp: 236-259.

Gaston, K. J. and Kunin, W. E. 1997 Concluding comments. In: The Biology of Rarity - Causes and Consequences of Rare-Common Differences (W. E. Kunin and K. J. Gaston, eds.), Chapman & Hall, London, Pp: 262-272.

Glazier, D. S. 1987n Toward a predictive theory of speciation: the ecology of isolate selection. Journal of Theoretical Biology 126:323-333.

Harrison, P. 1985 Seabirds: An Identification Guide. Edicions, Bar

Helbig, A. J. and celona.M. D 1999oe Molecular phylogeny of Palearctic-African Acrocephalus and Hippolais warblers (Aves: Sylviidae). Molecular Phylogenetics and Evolution 11b246-260.

Hughes, A. L. 1999el Differential human impact on the survival of genetically distinct avian lineages. Bird Conservation International 9i147-154.

Jablonski, D. 1986. Background and mass extinctions: the alternation of macroevolutionary regimes. Science 23:129-133.

Jablonski, D. 1987. Heritability at the species level: analysis of geographic ranges of Cretaceous mollusks. Science 238:360-363.

Johnson, K. P., McKinney, F. & Sorenson, M. D. 1999. Phylogenetic constraint on male parental care in the dabbling ducks. Proceedings of the Royal Society of London Series B-Biological Sciences 266:759-763.

Klicka, J. & Zink, R. M. 1997. The importance of recent ice ages in speciation: a failed paradigm. Science 277:1666-1669.

Kunin, W. E. 1997 Introduction: on the causes and consequences of rare-common differences. In: The Biology of Rarity - Causes and Consequences of Rare-Common Differences (W. E. Kunin and K. J. Gaston, eds.), Chapman & Hall, London, Pp.: 3-11.

Lawton, J. H. 1995 Population dynamic principles. In: Extinction Rates (J. H. Lawton and R. M. May), Oxford University Press, Oxford, Pp: 147-163.

Livezey, B. C. 1986 Phylogeny and historical biogeography of steamer-ducks (Anatidae: Tachyeres). Systematic Zoology 35:458-469.

Livezey, B. C. 1991. A phylogenetic analysis and classification of recent dabbling ducks (tribe Anatini) based on comparative morphology. Auk 108:471-507.

Livezey, B. C. 1995a. A phylogenetic analysis of the whistling and white-backed ducks (Anatidae: Dendrocygninae) using morphological characters. Annals of Carnegie Museum 64:65-97.

Livezey, B. C. 1995b. Phylogeny and comparative ecology of stiff-tailed ducks (Anatidae: Oxyurini). Wilson Bulletin 107:214-234.

Livezey, B. C. 1995c. Phylogeny and evolutionary ecology of modern seaducks (Anatidae: Mergini). Condor 97:233-255.

Livezey, B. C. 1996a. A phylogenetic analysis of geese and swans (Anseriformes: Anserinae), including selected fossil species. Systematic Biology 45:415-450.

Livezey, B. C. 1996b. A phylogenetic analysis of modern pochards (Anatidae: Aythyini). Auk 113:74-93.

Livezey, B. C. 1997. A phylogenetic analysis of modern sheldgeese and shelducks (Anatidae, Tadornini). Ibis 139:51-66.

Lockwood, J. L. 1999. Using taxonomy to predict success among introduced avifauna: relative importance of transport and establishment. Conservation Biology 13:560-567.

Lynch, J. D. 1989. The gauge of speciation: on the frequencies of modes of speciation. In: Speciation and its Consequences (D. Otte and J. A. Endler, eds.), Sinauer, Sunderland, Massachusetts.

Mace, G. M. and M. Kershaw. 1997. Extinction risk and rarity on an ecological timescale. In: The Biology of Rarity - Causes and Consequences of Rare-Common Differences (W. E. Kunin and K. J. Gaston, eds.), Chapman and Hall, London, Pp: 130-149.

Madge, S. and H. Burn 1988. Wildfowl: An Identification Guide to the Ducks, Geese and Swans of the World. Helm, London.

Manne, L. M., T.M. Brooks, and S.L. Pimm 1999. Relative risk of extinction of passerine birds on continents and islands. Nature 399:258-261.

Maurer, B. A. 1999. Untangling Ecological Complexity: The Macroscopic Perspective. University of Chicago Press, Chicago.

Mayr, E. 1963. Animal Species and Evolution. Harvard University Press, Cambridge, Massachusetts.

Mayr, E. 1982. Speciation and macroevolution. Evolution 36:1119-1132.

Mayr, E. 1988. Toward a New Philosophy of Biology: Observations of an Evolutionist. Harvard University Press, Cambridge, Massachusetts.

McKinney, M. L. 1997a. Extinction vulnerability and selectivity: combining ecological and paleontological views. Annual Review of Ecology and Systematics 28:495-516.

McKinney, M. L. 1997b. How do rare species avoid extinction? A paleontological view. In: The Biology of Rarity - Causes and Consequences of Rare-Common Differences (W. E. Kunin and K. J. Gaston, eds), Chapman & Hall, London, Pp: 110-129.

McKinney, M. L. 1998. Biodiversity dynamics: niche preemption and saturation in diversity equilibria. In: Biodiversity Dynamics (M. L. McKinney and J. Drake, eds.), Columbia University Press, Pp: 1-16.

4. Rarity and Phylogeny in Birds

McKinney, M. L. and J.L. Lockwood 1999. Biotic homogenization: a few winners replacing many losers in the next mass extinction. Trends in Ecology and Evolution 14:450-453.

McKitrick, M. C. 1993. Phylogenetic constraint in evolutionary theory: has it any explanatory power? Annual Review of Ecology and Systematics 24:307-330.

Mitchell, C. D. 1994. Trumpeter Swan (Cygnus buccinator). In: The Birds of North America No. 105 (A. Poole and F. Gill, eds.), The Academy of Natural Sciences, Philadelphia, The American Ornithologists' Union, Washington D.C.

Mooers, A. O. and P.H. Harvey 1994. Metabolic rate, generation time, and the rate of molecular evolution in birds. Molecular Phylogenetics and Evolution 3:344-350.

Myers, N. 1997. Mass extinction and evolution. Science 278:597-598.

Nee, S. and R.M. May 1997. Extinction and the loss of evolutionary history. Science 278:692-694.

Norris, R. D. 1991. Biased extinction and evolutionary trends. Paleobiology 17:388-399.

Nunn, G. B., J. Cooper, P. Jouventin, C.J.R. Robertson, and G.C. Robertson 1996. Evolutionary relationships among extant albatrosses (Procellariiformes: Diomedeidae) established from complete cytochrome-b gene sequences. Auk 113:784-801.

Orians, G. H. 1997. Evolved consequences of rarity. In: The Biology of Rarity - Causes and Consequences of Rare-Common Differences (W. E. Kunin and K. J. Gaston, eds.), Chapman & Hall, London, Pp: 190-208

Owen, M. and J.M. Black 1990. Waterfowl Ecology. Blackie, Glasgow & London.

Pagel, M. P., R.M. May, and A.R. Collie 1991. Ecological aspects of the geographic distribution and diversity of mammalian species. American Naturalist 137:791-815.

Peterson, A. T., J. Soberon, and V. Sanchez-Cordero 1999. Conservatism of ecological niches in evolutionary time. Science 285:1265-1267.

Price, T. D., A.J. Helbig and A.D. Richman 1997. Evolution of breeding distributions in the old world leaf warblers (genus Phylloscopus). Evolution 51:552-561.

Rabinowitz, D. 1981. Seven forms of rarity. In: The Biological Aspects of Rare Plant Conservation (H. Synge, ed.), Wiley, New York, Pp: 205-217.

Rao, P. S. R. S. and C.E. Heckler 1997. The three-fold nested random effects model. Journal of Statistical Planning and Inference 64:341-352.

Raup, D. M. 1991. Extinction: Bad Genes or Bad Luck? Oxford University Press, Oxford.

Reed, J. M. 1999. The role of behavior in recent avian extinctions and endangerments. Conservation Biology 13:232-241.

Reveal, J. L. 1981. The concepts of rarity and population threats in plant communities. In: Rare Plant Conservation (L. E. Morse and M. S. Henefin, eds.), The New York Botanical Garden, Bronx, New York, Pp: 41-46.

Ripley, S. D. and B.M. Beehler 1990. Patterns of speciation in Indian birds. Journal of Biogeography 17:639-648.

Rosenzweig, M. L. 1978. Geographical speciation: on range size and the probability of isolate formation. In: Proceedings of the Washington State University Conference on Biomathematics and Biostatistics (D. Wollkind, ed.), Washington State University, Washington, Pp: 172-194.

Rosenzweig, M. L. 1995. Species Diversity in Space and Time. Cambridge University Press, Cambridge.

Russell, G. J., T.M. Brooks, M.L. McKinney, and C.G. Anderson 1998. Present and future taxonomic selectivity in bird and mammal extinctions. Conservation Biology 12:1365-1376.

Schliewen, U. K., D. Tautz, and S. Pääbo 1994. Sympatric speciation suggested by monophyly of crater lake cichlids. Nature 386:629-632.

Sibley, C. G. and B.L. Monroe 1990. Distribution and Taxonomy of the Birds of the World. Yale University Press, New Haven.

Stanley, S. M. 1979. Macroevolution: Patterns and Process. W.H. Freeman, San Francisco.

Stattersfield, A. J., M.J. Crosby, A.J. Long, and D.C. Wege 1998. Endemic Bird Areas of the World. BirdLife International, Cambridge.

Tauber, C. and M.J. Tauber 1989. Sympatric speciation in insects: perception and perspective. In: Speciation and its Consequences (D. Otte and J. A. Endler, eds.),. Sinauer, Sunderland, Massachusetts, Pp: 307-344.

Taylor, C. M. and N.J. Gotelli 1994. The macroecology of Cyprinella: correlates of phylogeny, body size, and geographical range. American Naturalist 144:549-569.

Todd, F. S. 1996. Natural History of the Waterfowl. Ibis Publishing Company, Vista, California.

Tucker, G. M. and M.F. Heath 1994. Birds in Europe: Their Conservation Status. BirdLife International, Cambridge.

Vermeij, G. J. 1993. Biogeography of recently extinct marine species: implications for conservation. Conservation Biology 7:391-397.

Vitousek, P. M., H.A. Mooney, J. Lubchenco, and J.M. Melillo 1997. Human domination of Earth's ecosystems. Science 277:494-499.

Wagner, P. J. and D.H. Erwin 1995. Phylogenetic patterns as tests of speciation models. In: New Approaches to Speciation in the Fossil Record (D. H. Erwin and R. L. Anstey, eds.), Columbia University Press, New York, Pp: 87-122.

Williams, P. H. 1996. WORLDMAP 4 WINDOWS: software and help document 4.1. Privately Distributed, London.

Willis, J. C. 1922. Age and Area: A Study in Geographical Distribution and Origin of Species. Cambridge University Press, Cambridge.

Wilson, E. O. 1961. The nature of the taxon cycle in the Melanesian ant fauna. American Naturalist 95:169-193.

Chapter 5

Hybridization Between Native and Alien Plants and its Consequences

Curtis C. Daehler and Debbie A. Carino
Department of Botany, University of Hawai'i, Honolulu, HI 96822, USA

1. INTRODUCTION

With increasing frequency, humans have transported species around the globe, leading to geographic homogenization of formerly distinct communities. This loss of uniqueness among communities, and its consequences, have become major concerns for conservation biologists (Soule 1990). Competition and predation from alien species, as well as the spread of introduced diseases are well-recognized factors contributing the decline and disappearance of unique native species. More recently, genetic homogenization has been recognized as an additional consequence of alien species introductions. Genetic homogenization, at a global scale, results from hybridization between native and introduced species or populations. Rhymer and Simberloff (1996) reviewed cases of hybridization between native and alien species and populations, emphasizing examples and consequences in a wide variety of animal taxa. This chapter focuses on hybridization between native and alien plant species and its implications for plant conservation. We consider a species alien when it is known to have been introduced to a given region from a different region of the world.

Although we recognize that hybridization is sometimes defined as interbreeding between any two genetically distinct populations (e.g. Arnold 1997), for our purposes, we will focus on hybridization between two distinctly named species. Hybridization between genetically distinct populations of the same species must occur frequently, but we have little means of assessing this problem because few genetically distinct populations have been formally named or explicitly identified. After introducing the conservation issues associated with hybridization, we provide estimates of the scope of hybridization and hybridization potential between native and alien plants by analyzing the floras of the British Isles and the Hawaiian Islands. Finally, we conclude with a discussion of specific human-related factors that may have increased hybridization rates among plants over the past century.

2. CONSERVATION ISSUES

Natural hybridization between regionally sympatric species is a common phenomenon in the plant kingdom (Anderson 1949, Stace 1975, Ellstrand et al. 1996). Although hybridization may promote speciation in some cases (e.g. Anderson and Stebbins 1954, Stebbins 1959, Abbott 1992), hybridization can also threaten species with extinction (Levin et al. 1996). The conservation implications of hybridization over the long term will depend on the frequency of cross-pollination or hybridization and the fitness of the hybrids. Factors affecting the frequency of hybridization include flowering phenology, pollinator behavior, proximity and relative frequency of the parental species, and pollen-stigma interactions (reviewed in Arnold 1997). Factors affecting the fitness of hybrids include the particular species (and genotypes) involved, as well as the environment in which the hybrids are growing (Anderson 1948, Arnold and Hodges 1995). Previous reviews have emphasized the threats of hybridization to rare species and small populations (Rieseberg 1991, Ellstrand 1992, Levin et al 1996). While extinction by hybridization can certainly be an immediate threat to small populations, we will also emphasize that hybridization with an alien species can threaten even large native populations over the long term.

2.1 Alien pollen

Assuming that alien pollen is being deposited on the stigmas of a native species, there are several mechanisms that could threaten the native with extinction, even if viable hybrid seeds are never formed. First, the alien could fertilize the ovules, but hybrid seeds may be inviable due to abnormal endosperm development or abortion of hybrid embryos. Abortion of hybrid embryos, at the expense of pure seeds, reduces the native seed crop, resulting in population decline if the native population is seed limited. Although hybrid embryo abortion and lethal abnormal endosperm development in hybrids have been well documented in the plant breeding literature (Ladizinsky 1992), cases involving spontaneous native-alien hybridization have not been studied, so the extent of this problem is not known. One native-alien pair that could provide evidence for abortion of native ovules following pollination by an alien is the Canadian hawthorn, *Crataegus punctata*, and the European hawthorn, *C. monogyna*, now naturalized in Canada. When the Canadian species is experimentally pollinated by the European species, fewer than 3% of the Canadian hawthorn ovules form seeds (Wells and Phillips 1989). It is likely that the remaining 97% of ovules in the Canadian species are aborted after fertilization by the alien. The addition of native (Canadian hawthorn) pollen following pollination by the European species would clarify whether the native hawthorn ovules are aborted following fertilization by the alien or still accessible by the native pollen. Most cases of reduced native seed set due to abortion of hybrid seeds are likely to be missed by naturalists and ecologists. When viable hybrids are never observed in the field, it is often assumed that native-alien cross-fertilization is rare or does not occur. At the same time, when low seed set is observed in a rare native plant, factors such as inbreeding depression, pollen limitation, and resource limitation are typically called into play. If congeners are locally abundant, the effects of cross-pollination and the possibility of hybrid seed abortion should be seriously considered among factors that may be contributing to low seed set.

Even without fertilizing ovules, alien pollen could potentially decrease seed production in natives by occluding native stigmas, either at the stigma surface or in the style. Several studies have demonstrated that heterospecific or incompatible pollen can block conspecific pollen from accessing unfertilized ovules (Ockendon and Currah 1977, Galen and Gregory 1989, Petanidou et al. 1995 but see Armbruster and McGuire 1991, Gross 1996). Inhibition of conspecific pollen grains by heterospecific grains due to allelopathy has also been reported (Sukada and Jayachandra 1978). In general, large volumes of heterospecific or incompatible pollen are probably required to prevent conspecific pollen tubes from reaching ovules (e.g. Shore and Barrett 1984); therefore, stigma occlusion is likely to reduce seedset in a native species only when alien pollen is delivered frequently (e.g. from a large surrounding population) and in large volumes.

2.2 Hybrid fitness

Most documented examples of hybridization between native and alien species involved the production of viable hybrids. In these cases, the fitness of the hybrid will be an important factor in assessing risks to native species conservation. Reduced fitness (vigor or reproductive success) in hybrids has been termed outbreeding depression and may be attributed to disruption of coadaptive gene complexes (Templeton 1986, Ellstrand 1992). Differences in chromosome number or chromosome pairing problems between parental species can also reduce the fertility of hybrids (Arnold 1997). Similar to cases where hybrid embryos are aborted, at the expense of pure native progeny, a native may waste maternal resources on low fitness hybrid seeds that have little chance of survival. The eventual fate of the native population may be extinction as recruitment of pure native progeny becomes increasingly rare due to reduced seed output. The concept of low fitness hybrids can again be demonstrated with the hybridizing Canadian and European hawthorns (*C. punctata* and *C. monoyna*), although negative consequences for the native population have not yet been demonstrated. In Canada, natural hybrids are found sporadically, yet a hybrid swarm has not developed, probably because hybrids have reduced fitness relative to parental species (Wells and Phillips 1989). Backcrossing appears to be rare, and hybrids themselves have little chance of reproducing, as their seeds were found to be inviable (Wells and Phillips 1989). Although the native Canadian hawthorn wastes some resources on low-fitness hybrids, the Canadian hawthorn is a common species, and probably only a small fraction of seeds are hybrids. In this particular case, hybridization would seem to pose little threat to the common native. However, Arnold and Hodges (1995) and Arnold (1997) have emphasized the importance of fitness variability and rare events in the evolution and dynamics of hybrids. A rare genetic combination or arrangement in a hybrid could produce a high fitness hybrid or backcross that could rapidly spread. Furthermore, if the native Canadian hawthorn becomes rare in the future due to habitat destruction or other factors, then cross-pollination and hybridization with European hawthorn could become a serious conservation concern. In general, for cases where hybrids have low fitness, immediate threats to a native species are serious only if the alien fertilizes a very high proportion of the native's ovules.

When the fitness of hybrids approaches or exceeds that of the parental species, two factors, competition and genetic assimilation, can interact, potentially leading to the demise of even a common native species. Genetic assimilation occurs when the unique characters of one species are lost due to extensive hybridization and introgression (backcrossing of hybrids to the parental species). As the goal of many conservation programs is to preserve local uniqueness, genetic assimilation can be a serious problem (Rieseberg 1991, Ellstrand 1992). High to moderate fitness hybrids lead to a greater risk of genetic assimilation than low fitness hybrids because higher fitness hybrids are more likely to survive, spread, and backcross with the parental species than are weak hybrids with low fertility. With a larger pool of "genetically polluted" pollen and alien pollen available to fertilize native ovules, production of pure native seeds is expected to decline over time.

High fitness hybrids and introgressants may also replace pure native species through competition. Hybrids and introgressants between native species and invasive alien species might be predicted to have high fitness if they combine a native's adaptation to the local environment and an alien's invasive growth traits. Based on isozyme evidence, Streffeler et al. (1996) have suggested that the extremely invasive nature of European purple loosestrife (*Lythrum salicaria*) in the United States might be attributed hybridization with a native species, *L. alatum*, conferring local adaptations to the introgressants which morphologically resemble the alien. Similarly, Watanabe et al (1997a) suggested that local adaptations in the timing of germination in a native dandelion (*Taraxacum platycarpus*) could be transferred to hybrids formed with a vigorous alien *(T. officinale)*, resulting in very successful, invasive hybrids. Preferential dispersal of alien and hybrid seeds could also contribute to a decline in a native population. Vila and D'Antonio (1998) found alien *Carpobrotus edulis* and hybrid fruits were preferred by vertebrate dispersers over fruits of the putative native *C. chilensis*. Furthermore, gut passage enhanced germination in the hybrids but not in the natives (Vila and D'Antonio 1998).

High fitness native-alien species hybrids have been documented involving all plant major life forms from herbs to trees. In many cases, these hybrids directly threaten native species. For example, in the region of Manipur, India, construction of a hydroelectric dam has simultaneously decreased the habitat of the local wild rice, *Oryza rufipogon*, and increased the areas suitable for growing common rice (*O. sativa*). A hybrid swarm has now developed, and pure *O. rufipogon* has become exceedingly rare (Majumder et al. 1997). Extinction of a native rice species in Taiwan has also been reported following hybridization with cultivated *O. sativa* (Kiang et al. 1979). In another example, *Lantana camara*, a horticulturally grown shrub native to tropical America, has become naturalized in disturbed areas of Florida where it hybridizes with a rare native species, *L. depressa*. The long-term survival of *L. depressa* seems unlikely due to loss of genetic integrity and competition from vigorous hybrids (Sanders 1987). In Canada, populations of the native red mulberry (*Morus rubra*), a deciduous tree, are threatened with hybridization with the Asian white mulberry (*M. alba*) (Haber 1998). White mulberry has escaped from cultivation after being introduced to provide food for silkworms. Currently, there are more hybrid populations than pure native mulberry populations in Canada (Haber 1998). In some cases, introgression between native and alien species appears to have been extraordinarily extensive. In the Galapagos Islands, introgression between an endemic cotton, *Gossypium darwinii*, and cultivated cottons appears widespead, but because no pure populations of *G. darwinii* can be identified with certainty, the extent of introgression and degree of uniqueness of the original *G. darwinii* are now difficult to determine (Wendel and Percy 1990). In cases where hybrid swarms have developed, introgression may also result in the local demise of the pure alien species.

For example, the alien dandelion, *Taraxicum officinale,* had been considered naturalized at several sites in Japan, but recent allozyme analyses of individuals thought to be *T. officinale* indicated that most were probably introgressants with the native dandelion, *T. platycarpus* (Watanabe et al. 1997b). Pure alien *T. officinale* was confined to a relatively small, disturbed area. Local extinction of an alien species generally does not pose a conservation concern, since the pure alien stock remain in other parts of the world.

2.3 Importance of hybridization frequency and symmetry

Intuitively, the frequency of hybrid progeny produced by a native species is related to the relative size of the interfertile alien population (Ellstrand 1992, Levin et al. 1996). This has led to the implied assumption that a small alien population will pose little threat to a common native species, provided that the alien population does not increase, and hybrids are not fertile. It is important to recognize that the relative frequency of native versus alien plants does not necessarily correspond one-for-one to the relative frequency of native versus alien pollen on native stigmas. If the alien has much higher male fitness (pollen quantity or potency), then a rare alien could fertilize a disproportionately large number of native ovules, threatening even a relatively large native population. For example, in San Francisco Bay, alien Atlantic smooth cordgrass, *Spartina alterniflora*, currently has spread over a small area relative to the native California cordgrass, *S. foliosa,* but the alien produces over an order of magnitude more wind-dispersed pollen per area of habitat than the native (Anttila et al. 1998). Based on the composition of the pollen pool alone, even if the native is ten times more abundant than the alien is, the alien would be expected to father over half of the seeds produced by the native. Furthermore, in this case, pollen of the alien had higher germination rates on stigmas of the native than the native's own pollen, further increasing the proportion of hybrids seeds expected from native plants (Anttila et al. 1998). While the ovules of native California cordgrass were easily fertilized by pollen from the alien, the reverse was not true: application of native pollen to the alien's stigmas did not appear to produce seeds (Antilla et al. 1998). Such asymmetric hybridization, where the alien is more likely to be the father of a hybrid than the native, will lead to an increased threat of genetic assimilation of the native. In the case of the cordgrasses, hybrids are fertile and vigorous (Daehler and Strong 1997). All of these factors combined have the potential to quickly transform a large native population into a hybrid swarm.

5. Hybridization Between Native and Alien Plants

Similar asymmetric hybridization favoring the alien as the father has been reported between the European hawthorn, *Crataegus monogyna* and a native hawthorn (*C. douglassi*) in Oregon (Love and Feigen 1978). Asymmetric hybridization favoring fertilization of the native by alien pollen should be especially common in cases where the native is self-compatible and the alien is self-incompatible. Pollen from a self-incompatible species can often fertilize ovules of a self-compatible species, but the reverse is rare (Lewis and Crow 1958). In summary, asymmetric hybridization favoring seeds sired by the alien, high male fitness of an alien, and production of fertile hybrids can all increase the frequency of hybridization or rate of genetic assimilation, compared with expectations based simply on the relative abundance of the native and the alien species.

3. HOW FREQUENT IS NATIVE-ALIEN PLANT HYBRIDIZATION?

Most regional floras of the world lack comprehensive treatment of hybrids, so good estimates of the frequency of hybridization among native species are difficult to obtain (Ellstrand et al 1996). Native-alien hybridizations are even less well documented because many have arisen only in recent decades. One exception is the flora of the British Isles, where hybrid taxa of all kinds have been documented quite thoroughly (Stace 1991). We used primarily Stace (1991), but also Stace (1975) to assess the frequency of native-alien hybridization in the British Isles. Considering only angiosperm hybrids between species native to the British Isles and species alien to the British Isles, we identified 69 native-alien hybrids involving 52 alien and 55 native species (Table1). In total, about 4% of the 1295 alien species listed in Stace (1991) have hybridized with native species in the British Isles. However, if we consider only alien species with congeneric native species (712 species), nearly 7.5% of aliens have hybridized with natives. Conversely, 6% of native species have formed hybrids among those natives with an introduced congener (892 species). The aliens originate from a variety of regions, but the majority (67%) are native to some part of Europe (Table 1). While there appear to be many native-alien hybrids in certain plant families like Asteraceae, Poaceae and Rosacae (Table 1), this is expected because they are among the most speciose plant families in the world.

The native-alien hybrids do not appear to be randomly distributed among all British genera. A few genera, namely *Rumex*, *Epilobium*, and *Verbascum* contain large numbers of native-alien hybrids. In contrast, other large British genera like *Geranium*, *Trifolium*, and *Euphorbia*, each containing 10-

Table 1. Native-alien hybrids in the flora of the British Isles

Family	Hybrid[a]	Origin	Ref[b]
Apiaceae	Heracleum sphondylium x H. mantegazzianum	SW Asia	1
Asteraceae	Artemisia vulgaris x A. verlotiorum	China	1
Asteraceae	Erigeron acer x Conyza canadensis	N America	1
Asteraceae	Pilosella officinarum x P. aurantiaca	N + C Europe	1
Asteraceae	Senecio cineraria x S. erucifolius	Mediterranean	1
Asteraceae	Senecio cineraria x S. jacobaea	Mediterranean	1
Asteraceae	Senecio squalidus x S. vulgaris	S Europe	1
Asteraceae	Senecio vulgaris x S. vernalis	W Europe	1
Asteraceae	Solidago virgaurea x S. canadensis	N America	1
Betulaceae	Alnus glutinosa x A. incana	Europe	1
Betulaceae	Corylus avellana x C. maxima	SE Eur, SW Asia	1
Boraginaceae	Symphytum officianale x S. asperum	SW Asia	1
Brassicaceae	Capsella bursa-pastoris x C. rubella	S. Europe	1
Brassicaceae	Rorippa amphibia x R. austriaca	Europe	1
Brassicaceae	Rorippa sylvestris x R. austriaca	Europe	1
Caprifoliaceae	Viburnum lantana x V. rhytidophyllum	China	1
Caryophyllaceae	Cerastium arvense x C. tomentosum	Italy	1
Caryophyllaceae	Dianthus gratianopolitanus x D. caryophyllus	S Europe	1
Caryophyllaceae	Dianthus gratianopolitanus x D. caryophyllus x D. plumarius	S Europe	1
Caryophyllaceae	Dianthus gratianopolitanus x D. plumarius	SE Europe	1
Cheonopodiaceae	Chenopodium album x C. opulifolium	Europe	2
Cheonopodiaceae	Chenopodium album x C. berlandieri	N America	1
Cheonopodiaceae	Chenopodium album x C. suecicum	N Europe	2
Clusiaceae	Hypericum androsaemum x H. hircinum	Mediterranean	1
Convolvulaceae	Calystegia sepium x C. pulchra	?	1
Convolvulaceae	Calystegia sepium x C. silvatica	S Europe	1
Fagaceae	Quercus cerris x Q. robur	S. Europe	1
Lamiaceae	Mentha aquatica x M. spicata	?	1
Lamiaceae	Mentha arvensis x M. aquatica x M. spicata	?	1
Liliaceae	Hyacinthoides non-scritpa x H. hispanica	Spain, Portugal	1
Malvaceae	Malva neglecta x M. parviflora	S Europe	2
Onagraceae	Epilobium ciliatum x E. palustre	N America	1
Onagraceae	Epilobium hirsutum x E. ciliatum	N America	1
Onagraceae	Epilobium lanceolatum x E. ciliatum	N America	1
Onagraceae	Epilobium montanum x E. ciliatum	N America	1
Onagraceae	Epilobium obscurum x E. ciliatum	N America	1
Onagraceae	Epilobium parviflorum x E. ciliatum	N America	1
Onagraceae	Epilobium roseum x E. ciliatum	N America	1
Onagraceae	Epilobium tetragonum x E. ciliatum	N America	1

5. Hybridization Between Native and Alien Plants

Family	Hybrid[a]	Origin	Ref[b]
Poaceae	*Agrostis capillaris x A. castellana*	?	1
Poaceae	*Agrostis stolonifera x Polypogon viridis*	S Europe	1
Poaceae	*Festuca arundinaceae x Lolium multiflorum*	S Europe	1
Poaceae	*Festuca pratensis x. Lolium multiflorum*	S Europe	1
Poaceae	*Lolium perenne x L. multiflorum*	S Europe	1
Poaceae	*Spartina maritima x S alterniflora*	N. America	1
Polygonaceae	*Rumex confertus x R. crispus*	E Europe	1
Polygonaceae	*Rumex confertus x R. obtusifolius*	E Europe	1
Polygonaceae	*Rumex crispus x R. obovatus*	S America	2
Polygonaceae	*Rumex crispus x R. patientia*	Europe	2
Polygonaceae	*Rumex cristatus x R. crispus*	CS Europe	1
Polygonaceae	*Rumex cristatus x. R. obtusifolius*	E Europe	1
Polygonaceae	*Rumex frutescens x R. conglomeratus*	S America	1
Polygonaceae	*Rumex patienta x R. conglomeratus*	Europe	1
Polygonaceae	*Rumex patienta x R. obtusifolius*	Europe	1
Rosaceae	*Prunus spinosa x P. domestica*	SW Asia	1
Rosaceae	*Rosa multiflora x R. rubinginosa*	E Asia	1
Rosaceae	*Rosa rugosa x R. canina*	E Asia	1
Rosaceae	*Rubus idaeus x R. phoenicolasius*	East Asia	1
Rosaceae	*Sorbus aucuparia x S. intermedia*	Baltic region	1
Salicaceae	*Populus alba x P. tremula*	Europe	1
Saxifragaceae	*Saxifraga umbrosa x S. hirsuta*	Pyrenees	1
Saxifragaceae	*Saxifraga umbrosa x. spathularis*	Pyrenees	1
Scrophulariaceae	*Linaria purpurea x L. repens*	Italy	1
Scrophulariaceae	*Verbascum blattaria x V. nigrum*	Europe	1
Scrophulariaceae	*Verbascum phlomoides x V. thapsus*	Europe	1
Scrophulariaceae	*Verbascum pyramidatum x V. nigrum*	Caucasus	1
Scrophulariaceae	*Verbascum pyramidatum x V. thapsus*	Caucasus	1
Scrophulariaceae	*Verbascum thapsus x V. speciosum*	SE Europe	1
Scrophulariaceae	*Veronica longifolia x V. spicata*	N + C Europe	1

[a] Alien is listed in bold, hybrid nomenclature as given by Stace (1991) or Stace (1975)
[b] 1 = Stace (1991), 2= Stace (1975).

20 native species, have no native-alien hybrids. While our primary interest was not in determining which genera had significantly fewer hybrids than expected, we can demonstrate that native-alien hybrids were statistically under-represented among some genera using the family Polygonaceae as an example. Considering the total number of native and alien Polygonaceae in the British Isles, there are 218 possible hybrids. Of these, 4.13% actually have been recorded. The genus *Persicaria* (in the family Polygonaceae) contains 8 native and 9 alien species, or 72 possible hybrids, none of which have been observed. By calculating a binomial probability, we can determine that the finding of no native-alien hybrids in *Persicaria* is

statistically unlikely ($P = 0.048$). Similar calculations show statistical over- or under-representation of native-alien hybrids in other British genera (Daehler, unpublished).

Several hypotheses might explain the non-random distribution of the native-alien hybrids among British genera: 1) genera in Table 1 may show a higher propensity to hybridize (weak interspecific isolation) relative to the average genus, 2) the genera in Table 1 may be better studied than other genera, increasing the chances of discovering one or more native-alien hybrids, or 3) genera in Table 1 may contain more alien species than the average genus, leading to more possibilities for hybridization. We evaluated these hypotheses by pairing each genus in Table 1 with another genus from the same family containing an equal (or similar) number of native species, as listed in Stace (1991). We then compared our paired genera to determine if the genera in Table 1 contained more native-native hybrids than the random genera (hypothesis 1). We also compared the number of alien species in genera from Table 1 with the number of aliens in random genera (hypothesis 3). We used two criteria in selecting the genera for pairing with those in Table 1. First, we always chose a confamilial genus with the same or the closest number of native species to match each genus in Table 1. Second, we required each genus to include at least 1 alien species, since a genus without any alien species obviously cannot form native-alien hybrids. For some genera in Table 1, no comparable genera were available for pairing, and these genera were excluded from analysis (e.g. there was no confamilial genus that contains both native and alien species to match *Quercus*). Our methodology involving pairing confamilial genera does not attempt to address the question of whether families might differ in their propensity to form hybrids. Rather, we asked, after standardizing for possible differences among families (by pairing confamilial genera), do genera that contain native-alien hybrids differ from random genera?

In total, 24 genera from Table 1 could be paired with comparable genera without native-alien hybrids. The paired genera were compared statistically using Wilcoxon non-parametric paired sample tests (Zar 1996). As expected there was no significant difference between the mean number of native species belonging to genera in Table 1 and their paired genera (mean = 5.2 versus 4.6 native species, respectively, $Z = -0.88$, $P = 0.33$).

There was no significant difference between the number of native-native hybrids in genera from Table 1 versus random genera (mean = 2.4 versus 1.4 hybrids, respectively, $Z = -0.98$ $P = 0.33$), leading us to reject hypothesis 1. It appears that the genera in Table 1 do not have weaker inter-specific fertility barriers than other British genera that lack native-alien hybrids. For our comparison of native-native hybrids, we excluded the genus *Rosa* because centuries of artificial hybridizations by plant breeders have greatly inflated the number of native hybrids in this genus (Table1). Our finding of no significant difference in native-native hybrids between genera in Table 1 and random genera also leads us to reject hypothesis 2. If the genera in Table 1 had simply been better studied, then we would have expected more reports of native-native hybridization from those genera, which was not the case.

While propensity to form native hybrids does not seem to relate to propensity to form native-alien hybrids, the genera in Table 1 did contain significantly more introduced species than their paired genera (mean = 5.0 versus 3.2 alien species, respectively, $Z = -2.26$, $P = 0.02$). This indicates that after controlling for genus size (through pairing genera with similar numbers of native species), larger numbers of alien introductions increase the risk of a genus forming native-alien hybrids. This finding raises the question of whether there may be a "snowball effect" whereby increasing numbers of alien introductions leads to disproportionately higher numbers of native-alien hybrids. If we consider only the genera in Table 1, the number of hybrids per genus was not positively correlated with the percentage of possible native-alien hybrids per genus (number of native species * number of alien species) (Pearson correlation coefficient, $r = -0.15$, $P = 0.39$). This finding does not lend support to the "snowball effect" hypothesis, but we should point out that our test for a "snowball effect" was weak because most genera contained only one native-alien hybrid.

Of our three hypotheses, only hypothesis 3, that genera involved in native-alien hybridization contain more alien species, was statistically supported. Historical factors (specifically, the precise species that have been introduced, rather than their generic affinities) may also contribute to the distribution of native-alien hybrids among genera.

4. NATIVE-ALIEN HYBRIDIZATION POTENTIAL IN THE HAWAIIAN ISLANDS

As an isolated archipelago, the Hawaiian Islands have one of the highest rates of plant species endemism in the world. About 89% of the 956 flowering plant species native to the Hawaiian Islands are found nowhere else in the world (Wagner et al. 1990). Since there has been no comprehensive treatment of hybrids in the Hawaiian Islands, we used the most recent flora (Wagner et al 1990) to identify angiosperm plant genera containing both native species (usually endemic) and naturalized aliens in the Hawaiian Islands (Table 2). In assembling this list of potential native-

Table 2. Potential native-alien species hybrids in the Hawaiian Islands: Native flora of the Hawaiian Islands and naturalized congeners

Family	Native	E[1]	Alien	Orig[2]	Notes[3]
Amaranthaceae	*Amaranthus brownii*	1	*A. viridis*	widespread	
Aquifoliaceae	*Ilex anomala*	0	*I. aquifolium*	Europe	
Asteraceae	*Artemisia australis*	0	*A. vulgaris*	Eurasia	
Asteraceae	*Bidens spp.*	19	*Bidens spp.*	neotropics	P
Asteraceae	*Gnaphal. sandwicensium*	1	*G. purpureum*	N. Amer	P
Boraginaceae	*Heliotropium spp.*	0	*H. amplexicaule*	S. America	
Brassicaceae	*Lepidium arbuscula*	2	*L. virginicum*	N. Amer	
Campanulaceae	*Lobelia spp.*	13	*L. erinus*	S. Africa	
Caryophyllaceae	*Silene spp.*	5	*S. gallica*	Europe	
Chenopodiaceae	*Chenopodium oahuense*	1	*C. ambrosiodes*	Mexico	P
Convolvulaceae	*Ipomoea tuboides*	1	*I. trilobata*	W. Indies	
Convolvulaceae	*Jacquemontia ovalifolia*	0	*J. pentantha*	Mexico	
Cuscutaceae	*Cuscuta sandwichiana*	1	*C. campestris*	N. Amer	
Cyperaceae	*Cyperus spp.*	1	*Cyperus spp.*		
Cyperaceae	*Eleocharis obtusa*	0	*E. geniculata*	pantropical	
Cyperaceae	*Fimbristylis spp.*	1	*Fymbristylis spp.*		
Cyperaceae	*Mariscus spp.*	8	*M. meyenianus*	neotropics	
Cyperaceae	*Pycreus polystachyos*	0	*P. sanguinolentus*	Europe	
Cyperaceae	*Rhynchospora spp.*	0	*Rhynchospora spp.*		
Euphorbiaceae	*Chamaesyce spp.*	4	*Chamaesyce spp.*		
Euphorbiaceae	*Euphorbia haeleeleana*	1	*Euphorbia spp.*		
Euphorbiaceae	*Phyllanthus distichus*	1	*P. debilis*	India	
Fabaceae	*Acacia koa*	1	*Acacia spp.*		
Fabaceae	*Caesalpinia kawaiensis*	1	*C. decapetala*	Asia	
Fabaceae	*Canavalia spp.*	6	*C. cathartica*	paleotrop	
Fabaceae	*Senna gaudichaudii*	0	*Senna spp.*		
Fabaceae	*Sesbania tomentosa*	1	*S. sesban*	Asia	
Fabaceae	*Vicia menziesii*	1	*V. sativa*	Europe	
Gentianaceae	*Centaurium sebaeoides*	1	*C. erythraea*	Europe	
Geraniaceae	*Geranium spp.*	5	*Geranium spp.*		

5. Hybridization Between Native and Alien Plants

Family	Native	E[1]	Alien	Orig[2]	Notes[3]
Iridaceae	*Sisyrinchium acre*	1	*S. exile*	S. America	
Malvaceae	*Abutilon spp.*	2	*Abutilon spp.*		
Malvaceae	*Gossypium tomentosum*	1	*Gossypium spp.*		N, A
Malvaceae	*Hibiscus spp.*	4	*Hibiscus spp.*	Asia	A
Malvaceae	*Sida fallax*	0	*Sida spp.*		P
Myrtaceae	*Eugenia reinwardtiana*	0	*E. uniflora*	S. America	
Myrtaceae	*Syzgium sandiwcensis*	1	*S. cumini*	India	
Nyctaginaceae	*Boerhavia spp.*	1	*B. coccinea*	Caribbean	
Papaveraceae	*Argemone glauca*	1	*A. mexicana*	Mexico	
Phytoloccaceae	*Phytolacca sandwicensis*	0	*P. octandra*	neotropics	
Piperaceae	*Peperomia spp.*	23	*P. pellucida*	neotropics	
Pittosporaceaee	*Pittosporum spp.*	8	*Pittosporum spp.*		
Plantaginaceae	*Plantago spp.*	3	*Plantago spp.*		
Poaceae	*Agrostis spp.*	1	*Agrostis spp.*		
Poaceae	*Cenchrus agrimoniodes*	1	*Cenchrus spp.*		
Poaceae	*Eragrostis spp.*	5	*Eragrostis spp.*		
Poaceae	*Panicum spp.*	11	*Panicum spp.*		
Poaceae	*Poa spp.*	2	*Poa spp.*		
Poaceae	*Sporobolus virginicus*	0	*Sporobolus spp.*		
Polygonaceae	*Rumex spp.*	3	*Rumex spp.*		
Portulacaceae	*Portulaca spp.*	3	*Portulaca spp.*		N
Ranunculaceae	*Ranunculus spp.*	2	*Ranunculus spp.*		
Rosaceae	*Fragraria chiloensis*	0	*Fragraria vesca*	widespread	
Rosaceae	*Rubus spp.*	3	*Rubus spp.*	Asia	N, A
Rubiaceae	*Hedyotis spp.*	20	*Hedyotis spp.*		
Rubiaceae	*Morinda trimera*	1	*M. citrifolia*	Asia	
Solanaceae	*Solanum spp.*	3	*Solanum spp.*		P
Urticaceae	*Pilea peploides*	0	*P. microphylla*	S. America	
Zygophyllaceae	*Tribulus cistoides*	0	*T. terrestris*	Europe	

[1] Number of species endemic to the Hawaiian Islands
[2] If genus contains multiple aliens of different origins, origin is not listed.
[3] N= natural hybrids confirmed from field (see text), A= artificial hybrids produced (Neal 1965),
P= putative hybrids identified from field (personal communication with local botanists).

alien hybrids, we included only cases where a naturalized alien species grows in a similar habitat and at a similar elevation to a native congener. Although in many cases, chromosome numbers between aliens and natives were the same, we did not consider chromosome numbers in generating

Table 2. Even large differences between species in chromosome numbers do not necessarily preclude hybridization (Stace 1975, Sanders 1987, Ashton and Abbott 1990).

Because native-alien hybridization has been so little studied in the Hawaiian Islands (and in most regions of the world for that matter), we have very little solid evidence for natural hybridization among most of the taxa in Table 2. Three cases of spontaneous native-alien hybridization have been reasonably well documented in the Hawaiian Islands, two within the past decade. Randell et al. (1998) used polymorphic DNA markers to identify natural hybrids between alien *R. rosifolius* and endemic *Rubus hawaiensis* (frequently cited as having lost its prickles due to evolution in the absence of large herbivores; it can still produce small prickles). Hybrids were documented on the island of Maui, but *R. rosifolius*, introduced from Jamaica, is a widespread weed in the Hawaiian islands (Wagner et al. 1990), and additional hybrid populations may occur throughout the islands. In the second case, Kim and Carr (1990) used chromosome counts of a field specimen and morphological comparisons with known hybrids in the lab to infer that native *Portulaca lutea* ($2n = 40$) had hybridized with alien *P. oleracea* ($2n=54$), forming a stable allopolyploid in the field. Other native and alien *Portulaca* species were also capable of hybridizing (Kim and Carr 1990). The third confirmed case of native-alien hybridization involves the endemic Hawaiian cotton (*Gossypium tomentosum*) which has hybridized with introduced *G. barbadense*, as determined by detailed morphological studies of neighboring populations (Stephens 1964). Fertility of these hybrids is high (deJoode and Wendel 1992). In Table 2 we have indicated other cases where artificial hybrids have been produced and where putative hybrids have been observed in the field, based on our personal communications with local botanists.

A total of 59 genera consisting of 176 endemic flowering plant species in the Hawaiian Islands are potentially threatened with hybridization by alien congeners (Table 2). This figure will be an overestimate if some of the native-alien congeners are intersterile (but see *Alien pollen* above). At the same time, there is good reason to believe our estimate of the threats of native-alien hybridization could be an underestimate. We did not consider the possibility of intergeneric hybrids in compiling Table 2. For example, the endemic Hawaiian genus *Lipochaeta* (16 extant species; Asteraceae) is interfertile with a widespread alien, *Wedelia trilobata* (Rabakonandrianina and Carr 1981). In compiling Table 2, we also did not consider the possibility of hybridization between native species and non-naturalized aliens that have been planted widely for horticulture or as crops. If these

5. Hybridization Between Native and Alien Plants

Table 3. Potential native-alien hybrids in the Hawaiian Islands: Native flora of the Hawaiian Islands and horticulturally grown (but largely non-naturalized) congeners

Family	Native	Endemics	Alien	Origin[1]
Apocynaceae	*Ochrosia spp.*	4	*O. elliptica*	New Caledonia
Araceae	*Prichardia spp.*	19	*Prichardia spp.*	Fiji
Capparidaceae	*Capparis sandwichiana*	1	*C. petiolaris*	S. America
Ebenaceae	*Diospyros spp.*	1	*Diospyros spp.*	
Euphorbiaceae	*Antidesma spp.*	2	*A. bunius*	Asia
Euphorbiaceae	*Rhus sandwicensis*	1	*Rhus spp.*	
Fabaceae	*Erythrina sandwicensis*	1	*Erythrina spp.*	
Fabaceae	*Mucuna gigantea*	0	*Mucuna spp.*	
Fabaceae	*Sophora chrysophylla*	1	*Sophora spp.*	
Fabaceae	*Strongylodon ruber*	1	*S. marcrobotrys*	Philippines
Fabaceae	*Vigna spp.*	1	*Vigna spp.*	
Moraceae	*Streblus pendulinus*	0	*S asper*	S. Asia
Myrtaceae	*Metrosideros spp.*	4	*M. tomentosa*	New Zealand
Nyctaginaceae	*Pisonia spp.*	2	*Pisonia alba*	Asia
Plumbaginaceae	*Plumbago zeylandica*	0	*Plumbago spp.*	
Primulaceae	*Lysimachia spp.*	9	*Lysimachia*	
Rhamnaceae	*Alphitonia ponderosa*	1	*A. zizyphoides*	Polynesia
Rubiaceae	*Coprosma spp.*	13	*C. baueri*	New Zealand
Rubiaceae	*Gardenia spp.*	3	*Gardenia spp.*	
Rutaceae	*Zanthoxylum spp.*	4	*Z. piperitum*	China
Santalaceae	*Santalum spp.*	4	*Santalum album*	India
Sapindaceae	*Sapindus spp.*	1	*S. mukorossii*	China
Sapotaceae	*Pouteria sandwicensis*	1	*Pouteria spp.*	
Solanceae	*Lycium sandwicense*	0	*L. chinense*	Asia
Urticaceae	*Boehmeria grandis*	1	*B. nivea*	S. Asia
Verbenaceae	*Vitex rotundifolia*	0	*Vitex spp.*	
Violaceae	*Viola spp.*	7	*V. odorata*	Eurasia

[1] If multiple aliens of different origins, origin is not listed

alien plants are also considered (Table 3), an additional 27 genera, consisting of 82 endemic species, are at risk from native-alien hybridization. Although non-naturalized alien plants are widely assumed to pose little risk to native plants, since they do not directly compete with natives in natural areas, risks of hybridization and introgression should be considered. Biotic pollinators can disperse pollen surprisingly long distances (Broyles and Wyatt 1991, Boshier et al. 1995, Dawson et al. 1997), and the rare production of even a single fertile hybrid could pose an immediate threat to a highly localized endemic species. Many populations of endangered Hawaiian plants number fewer than 20 individuals in the wild.

If we apply the rate of native-alien species hybridization from the British Isles (6% of natives have formed hybrids) to the flora of the Hawaiian Islands, then we would expect about 15 endemic species to hybridize with alien species. There is reason to suspect that the threat of native-alien hybridization in the Hawaiian Islands is higher than in the British Isles. While many speciose genera in the Hawaiian Islands contain an astounding array of life forms and ecological adaptations (e.g. *Bidens* and *Hedyotis* listed in Table 2), they also have weak interspecific fertility barriers (Carlquist 1974, Ganders and Nagata 1985, Wagner et al. 1990). If a widespread alien can hybridize with one species in one of these genera, it is also likely interfertile with the other species from that genus, substantially boosting the total number of native-alien hybrids that are likely to develop. Similar interfertility among endemic species in other island floras, combined with small population sizes, may make island floras more susceptible to extinction by hybridization than mainland floras, independent of whether aliens are involved (Levin et al. 1996).

5. RECENT GLOBAL CHANGES THAT MAY AFFECT RATES OF NATIVE-ALIEN HYBRIDIZATION

5.1 Disturbance and hybridization

Humans in one way or another have disturbed almost all areas of the Earth. Disturbed habitats have long been recognized as havens for hybrids (Wiegand 1935, Anderson 1948). In general, hybrids have been proposed to thrive in disturbed habitats because disturbance frees resources and creates new, sometimes intermediate habitats that are especially suitable for hybrids (Anderson 1948). Each parental species is often thought to be adapted to a unique environment, and the inability of intermediate hybrids to compete with the pure parental species in the undisturbed, parental environments also explains why hybrids often disappear in the absence of persistent disturbance (Anderson 1948). The importance of competition from parental species in limiting the spread of hybrids is questionable when an alien species is involved in the hybridization. The alien parent usually does not have adaptations specific to the local biotic and abiotic environment in the invaded region, so there may be no specific local adaptations that are disrupted in the hybrids formed with a native species. In contrast, hybrids may inherit aggressive growth characteristics of the alien and some local adaptations of the native, allowing them to rapidly spread (see *Hybrid Fitness*).

A second explanation for the abundance of hybrids in disturbed habitats is that disturbance can bring two parental species into close proximity, increasing rates of cross-pollination and hybridization. Two species that would normally grow in different habitats may grow together in a disturbed habitat. Aliens, like hybrids, are often more common in habitats affected by human disturbance (Allan 1936, Fox and Fox 1986), and for this reason alone, native-alien hybrids might be expected at higher frequencies in disturbed habitats.

5.2   Decline of native pollinators

Native pollinators are on the decline in many parts of the world due to habitat destruction, parasites, disease, and competition from introduced pollinators (Allen-Wardell et al. 1998). In contrast to the widespread decline of native pollinators, European honeybees (*Apis mellifera*) and other generalist alien bees originally introduced for crop pollination are now common in natural areas (e.g. Roubik 1986, Sugden and Pyke 1991, Paton 1993). The replacement of relatively specialized oligolectic native bees (bees that visit a few species of flowers) with generalist alien bees has the potential to increase the rates of plant hybridization. To date, there have been no scientific studies to specifically address whether generalist alien pollinators increase rates of native-alien plant hybridization, so we a can only speculate about the implications of replacing native pollinators with alien generalists. Huryn (1997) claims that pollen examined from the bodies of individual honeybees has only rarely revealed "mixed" pollen loads, which would imply that increased hybridization due to alien honeybees is unlikely. Unfortunately, in most cases, plant species within genera cannot be distinguished based on their pollen morphology (Faegri and Iversen 1989), making monospecific versus mixed pollen loads impossible to determine when congeneric plant species are located within a foraging area. Generalist honeybees may sequentially visit congeneric plant species that would otherwise be serviced by different pollinators. It is even conceivable that native pollinators might avoid flowers of certain alien plant species (preventing cross-pollination between a native and alien congener), while an introduced generalist pollinator may transfer pollen from the alien to the native.

The large foraging areas used by honeybee colonies could also promote increased hybridization between physically isolated native and alien species. A single colony of European honeybees may forage over an area of 48 hectares (Proctor et al. 1996), and individual honeybees can forage across distances of more than 3000 m (Stabentheiner 1996). Native solitary bees are likely to exhibit more localized foraging. Although the honeybee's generalist nature often does not make it an efficient pollinator (Westercamp 1991, Keys et al. 1995, Gross and Mackay 1998), artificially high honeybee densities supported by human-maintained colonies may make up for low pollen transfer efficiencies (Proctor et al. 1996), allowing some native-alien pollen transfer that would otherwise not take place.

5.3   Crop and horticulture hybrids

The potential for crop plants to form hybrids with weedy relatives has long been recognized (Small 1984). More recently, with the release of genetically engineered crops, much attention has been focused on the frequency and risks of cross-pollination between crops and their wild relatives (e.g. Ellstrand and Hoffman 1990, Parker and Kareiva 1996). The common practice of creating artificial hybrids in the horticulture and plant breeding industries poses additional risks to native plants, since these artificial hybrids may form a bridge for gene transfer between two formerly intersterile species. Humans have devised numerous techniques, from chemical treatment of stigmas (Willing and Pryor 1976) to embryo culture (Ladizinsky 1992) to somatic cell fusion (Sink et al. 1992) in order to overcome interspecific incompatibilities. For example, Ladizinsky (1992) lists over 100 crop hybrids obtained from embryo culture between otherwise incompatible species. Thousands of interspecific hybrids have also been developed for horticulture (Trehane 1989). Some of these involve hybridization of species derived from different geographic regions; numerous examples from gardens of Europe are listed in Trehane (1989). Unfortunately, for most horticultural hybrids, the breeder has not divulged the parental species, making analyses of hybrid breeding trends difficult. In many cases, artificial hybrids are partially or fully fertile and may be capable of backcrossing to one or more of the parental species. Even if a hybrid is initially sterile, chromosomal rearrangements over time may restore fertility and cross-compatibility to parental species (e.g. Hauber and Bloom 1983), leading to the potential for natural introgression among distantly related and normally intersterile species. Furthermore, based on experimental evidence, Harlan and DeWet (1963) have suggested that hybrids derived from wide crosses are themselves more capable of crossing with additional species, which could lead to a "snowball" effect, where a hybrid consumes the genetic distinctness of many species.

6. CONCLUSIONS

Increasing rates of plant introductions world-wide have led to an increasing recognition of the threats of genetic homogenization via hybridization between native and alien plants. The distinctness of endemic species, long valued by conservationists, can be lost though introgression with widespread alien species. Over the long-term, even large native populations may be threatened. Much introgression is likely to be cryptic and overlooked because introgressants resemble parental species (Anderson 1948). Increased use of molecular tools will likely reveal previously unsuspected hybridization and introgression (Rieseberg and Wendel 1992). While some authors have emphasized the creative force of hybridization in the evolution of novelty and speciation (e.g. Stebbins 1959, Knobloch 1972), hybridization associated with current high rates of species introductions and continued immigration of aliens is likely to lead to genetic assimilation or homogenization rather than increased regional differentiation.

REFERENCES

Allan, H. H. 1936. Indigene versus alien in the New Zealand plant world. Ecology 17:187-192.
Allen-Wardell, G., P.Bernhardt, R.Bitner, A. Burquez, S. Buchmann, et. al. 1998. The potential consequences of pollinator declines on the conservation of biodiversity and stability of food crops. Conservation Biology 12:8-17.
Anderson, E. 1948. Hybridization of the habitat. Evolution 2:1-9.
Anderson, E. 1949. Introgressive hybridization. New York: John Wiley and Sons.
Anderson, E., and G. L. Stebbins. 1954. Hybridization as an evolutionary stimulus. Evolution 8:378-388.
Anttila, C. K., C. C. Daehler, N. E. Rank, and D. R. Strong. 1998. Greater male fitness of a rare invader (Spartina alterniflora, Poaceae) threatens a common native (Spartina foliosa) with hybridization. American Journal of Botany 85:1597-1601.
Armbruster, W. S., and A. D. McGuire. 1991. Experimental assessment of reproductive interactions between sympatric Aster and Erigeron (Asteraceae) in interior Alaska. American Journal of Botany 78:1449-1457.
Arnold, M. L. 1997. Natural hybridization and evolution. Oxford University Press, New York.
Arnold, M. L., and S. A. Hodges. 1995. Are natural hybrids fit or unfit relative to their parents? Trends in Ecology and Evolution 10:67-71.

Ashton, P. A., and R. J. Abbott. 1992. Multiple origins and genetic diversity in the newly arisen allopolyploid species, Senecio cambrensis Rosse (Compositae). Heredity 68:25-32.

Carlquist, S. 1974. Island Biology. Columbia University Press, New York.

Daehler, C. C., and D. R. Strong. 1997. Hybridization between introduced smooth cordgrass (*Spartina alterniflora*; Poaceae) and native California cordgrass (*S. foliosa*) in San Francisco Bay, California, USA. American Journal of Botany 84:607-611.

DeJoode, D. R., and J. F. Wendel. 1992. Genetic diversity and origin of the Hawaiian Island cotton, Gossypium tomentosum. American Journal of Botany 79:1311-1319.

Ellstrand, N. C. 1992. Gene flow by pollen: Implications for plant conservation genetics. Oikos 63:77-86.

Ellstrand, N. C., and C. A. Hoffman. 1990. Hybridization as an avenue of escape for engineered genes. Bioscience 40:438-442.

Ellstrand, N. C., R. Whitkus, and L. H. Rieseberg. 1996. Distribution of spontaneous plant hybrids. Proceedings of the National Academy of Science, USA 93:5090-5093.

Faegri, K., and J. Iversen. 1989. Textbook of pollen analysis. John Wiley and Sons, New York.

Fox, M. D., and B. J. Fox. 1986. The susceptibility of natural communities to invasion. In: Ecology of biological invasions: An Australian perspective (R. H. Groves and J. J. Burden, eds.), Cambridge University Press, Canberra, Pp: 57-66.

Galen, C., and T. Gregory. 1989. Interspecific pollen transfer as a mechanism of competition: Consequences of foreign pollen contamination for seed set in the alpine wildflower, Polemonium viscosum. Oecologia 81:120-123.

Ganders, F. R., and K. M. Nagata. 1984. The role of hybridization in the evolution of Bidens on the Hawaiian Islands. In: Plant Biosystematics (W. F. Grant ed.), Academic Press, New York, Pp: 179-194.

Gross, C. L. 1996. Is resource overlap disadvantageous to three sympatric legumes? Australian Journal of Botany 21:133-143.

Gross, C. L., and D. Mackay. 1998. Honeybees reduce fitness in the pioneer shrub Melastoma affine (Melastomataceae). Biological Conservation 86:169-178.

Haber, E. 1998. Impact of invasive plants on species and habitats at risk in Canada. Canadian Wildlife Service Report, Ottawa. (available at http://infoweb.magi.com/~ehaber/impact.html)

Harlan, J. R., and J. M. J. deWet. 1963. The compilospecies concept. Evolution 17:497-501.

Hauber, D. P., and W. L. Bloom. 1983. Stability of a chromosomal hybrid zone in the Clarkia nitens and Clarkia speciosa ssp. polyantha complex (Onagraceae). American Journal of Botany 70:1454-1459.

Keys, R. N., S. L. Buchmann, and S. E. Smith. 1995. Pollination effectiveness and pollination efficiency of insects foraging Prosopis velutina in south-eastern Arizona. Journal of Applied Ecology 32:519-527.

Kiang, Y. T., J. Antonovics, and L. Wu. 1979. The extinction of wild rice (*Oryza perenis formsana*) in Taiwan. Journal of Asian Ecology 1:1-9.

Kim, I., and G. D. Carr. 1990. Cytogenetics and hybridization of Portulaca in Hawaii. Systematic Botany 15:370-377.

Knobloch, E. W. 1972. Intergeneric hybridization in flowering plants. Taxon 21:97-103.

Ladizinsky, G. 1992. Crossability relations. In: Distant hybridization of crop plants (G. Kalloo and J. B. Chowdhury eds.), Springer-Verlag, Berlin, Pp: 15-31.

Levin, D. A., J. Francisco-Ortega, and R. K. Jansen. 1996. Hybridization and the extinction of rare plant species. Conservation Biology 10:10-16.

Lewis, D. A., and L. K. Crowe. 1958. Unilateral interspecific incompatibility in flowering plants. Heredity 12:233-256.

5. Hybridization Between Native and Alien Plants

Love, R., and M. Feigen. 1978. Interspecific hybridization between native and naturalized Crataegus (Rosaeae) in western Oregon. Madrono 25:211-217.

Majumder, N. D., T. Ram, and A. C. Sharma. 1997. Cytological and morphological variations in hybrid swarms and introgressed populations of interspecific hybrids (*Oryza rufipogon* Griff. x *Oryza sativa* L.) and its impact on evolution of intermediate types. Euphytica 94:295-302.

Ockendon, D. J., and L. Currah. 1977. Self-pollen reduces the number of cross-pollen tubes in the styles of Brassica oleracaea L. New Phytologist 78:675-680.

Parker, I. M., and P. Kareiva. 1996. Assessing the risks of invasion for genetically engineered plants: acceptable evidence and reasonable doubt. Biological Conservation 78:193-203.

Paton, D. C. 1993. Honeybees in the Australian environment. Bioscience 43:95-103.

Petanidou, T., J.C. Den-Nijs, and J.G.B. Oostermeijer. 1995. Pollination ecology and constraints on seed set of the rare perennial *Gentiana cruciata* L. in The Netherlands. Acta Botanica Neerlandica 44:55-74.

Proctor, M., P. Yeo, and A. Lack. 1996. The natural history of pollination. Timber Press, Portland.

Rabakonandrianina, E., and G. D. Carr. 1981. Intergeneric hybridization, induced polyploidy, and the origin of the Hawaiian endemic Lipochaeta from Wedelia (Compositae). American Journal of Botany 68:206-215.

Randell, R. A., S. O. Grose, D. E. Gardner, and C. W. Morden. 1998. Hybridization among endemic and naturalized species of Rubus (Rosaceae) in the Hawaiian Islands. Paper presented at joint meeting of American Society of Naturalists/Society of Systematic Biologists/Society for the Study of Evolution, University of British Columbia - Vancouver, B.C., Canada

Rieseberg, L. H. 1991. Hybridization in rare plants: Insights from case studies in Cercocarpus and Helianthus. In: Genetics and conservation of rare plants (D. A. Falk and K. E. Holsinger eds.), Oxford University Press, Oxford, Pp: 171-181

Rieseberg, L. H., and J. F. Wendel. 1993. Introgression and its consequences in plants. In R.G. Harrison ed., Hybrid zones and the evolutionary process, pp. 70-109. New York: Oxford University Press.

Rhymer, J. M., and D. S. Simberloff. 1996. Extinction by hybridization and introgression. Annual Review of Ecology and Systematics 27:83-109.

Roubik, D. W. 1986. Sporadic food competition with the African honeybee: projected impact on neotropical social bees. Journal of Tropical Ecology 2:97-111.

Sanders, R. W. 1987. Identity of *Lantana depressa* and *L. ovatifolia* (Verbenaceae) of Florida and the Bahamas. Systematic Botany 12: 4-60.

Shore, J. S., and S. C. H. Barrett. 1984. The effect of pollination intensity and incompatible pollen on seed set in *Turnera ulmifolia* (Turneraceae). Canadian Journal of Botany 62:1298-1303.

Sink, K. C., R. K. Jain, and J. B. Chowdhury. 1992. Distant hybridization in crop plants. In: Distant hybridization of crop plants (G. Kalloo and J. B. Cowdhury eds.), Springer-Verlag, Berlin, pp. 168-198.

Small, E. 1984. Hybridization in the domesticated-weed-wild complex. In: Plant Biosystematics (W.F. Grant ed.), Academic Press, New York, Pp: 195-210.

Soule, M. E. 1990. The onslaught of alien species, and other challenges in the coming decades. Conservation Biology 4:233-239.

Stabentheiner, A. 1996. Effect of foraging distance on the thermal behaviour of honeybees during dancing, walking and trophallaxis. Ethology 102:360-370.

Stace, C. A. 1975. Hybridization and the flora of the British Isles. Cambridge University Press, Cambridge.

Stace, C. A. 1991. New Flora of the British Isles. Cambridge University Press, Cambridge.

Stebbins, G. L. 1959. The role of hybridization in evolution. Proceedings of the American Philosophical Society 103:231-251.

Streffeler, M. S., E. Darmo, R. L. Becker, and E. J. Katovich. 1996. Isozyme characterization of genetic diversity in Minnesota populations of purple loosestrife, *Lythrum salicaria* (Lythraceae). American Journal of Botany 83:265-273.

Stephens, S. G. 1964. Native Hawaiian cotton (*Gossypium tomentosum* Nutt.). Pacific Science 18:385-398.

Sukada, D. K., and Jayachandra. 1980. Pollen allelopathy: A new phenomenon. New Phytologist 84:739-746.

Trehane, P. 1989. Index Hortensis Volume 1: Perennials. Quarterjack Publishing, Wimorne.

Wagner, W. L., D. R. Herbst, and S. H. Sohmer. 1990. Manual of the flowering plants of Hawaii, volume 1. University of Hawaii Press. Honolulu.

Watanabe, M., M. Ogawa, S. Serizawa, M. Kanzaki, and T. Yamakura. 1997a. Intrusion of hybridized alien dandelions into the native dandeion's habitats. Acta Phytotaxonomica et Geobotanica 48:73-78.

Watanabe, M., Y. Maruyama, and S. Serizawa. 1997b. Hybridization between native and alien dandelions in the western Tokai district (1) Frequency and morphological characters of the hybrid between Taraxacum platycarpum and T. officinale. Journal of Japanese Botany 72:51-55.

Westercamp, C. 1991. Honeybees are poor pollinators: Why? Plant Systematics and Evolution 177:71-75.

Wendel, J. F., and R. G. Percy. 1990. Allozyme diversity and introgression in the Galapagos Islands (Pacific Ocean) endemic Gossypium darwinii and its relationship to continental Gossypium barbadense. Biochemical Systematics and Ecology 8:517-528.

Wiegand, K. M. 1935. A taxonomist's experience with hybrids in the wild. Science 81:161-166.

Willing, R. R., and L. D. Pryor. 1976. Interspecific hybridization in poplar. Theoretical and Applied Genetics 47:141-151.

Zar, J. H. 1996. Biostatistical Analysis, Third Edition. Prentice-Hall, Upper Saddle River, New Jersey.

Chapter 6

Taxonomic Selectivity in Surviving Introduced Insects in the United States

Diego P. Vázquez and Daniel Simberloff
Department of Ecology and Evolutionary Biology, University of Tennessee, Knoxville, TN 37996-1610, U.S.A.

1. INTRODUCTION

Although ecologists have understood the importance of invasions since the publication of Elton's book (1958), the factors important to the survival and impact of introduced species are still poorly understood. Several factors can determine whether a species arrives and thrives in a new location. Firstly, the opportunity for colonisation is important. Obviously, a species may have all the traits that would predispose it to prosper in a particular location, but, if it cannot get there, no invasion will occur. Clearly, some species will be introduced more often than others will. For example, the fact that there has been historically more traffic between Europe and North America than between Africa and North America gave European species more opportunities for spread to North America (Sailer 1983).

Secondly, the abiotic and biotic characteristics of the colonised habitat may be important. A tropical species might be ill suited to colonise high latitude areas with colder climates. On the other hand, less extreme habitat differences would probably make a relationship between habitat characteristics and the probability of survival less clear-cut. Biotic interactions between the invader and the native biota might determine whether the invader would thrive — e.g., the biotic resistance hypothesis (Simberloff 1986). As another example, a host-specific insect cannot survive long if its host is not already present. Lonsdale (1999) recently has shown that habitat characteristics (number of native species, whether the site was on the mainland or an island, and whether or not it was a nature reserve) explained much of the variation in the number of invasive plant species in

184 sites around the world. To our knowledge, no similar analysis has been conducted for insects, probably because the necessary data still have to be gathered.

Thirdly, the biological traits of invaders may also be important in determining their success. Historically, many researchers have attempted generalisations concerning the characteristics of invaders. For example, Baker (1965) hypothesized the "ideal weed" to be a species able to germinate in a wide range of conditions, grow quickly, flower early, self-fertilize, produce many seeds that disperse widely, reproduce vegetatively, and compete well. However, Baker's hypothesis is not supported by existing data. For example, Williamson and Fitter (1996) tested these and other life history characteristics with the Ecological Flora Database, finding weakly significant differences between good and poor invaders for only three of fourteen life history and reproductive characteristics (age at first flowering, type of pollen vector, and decline). Lawton and Brown (1986) reached a similar conclusion in analyzing the characteristics of insect invaders: only a combination of the population's intrinsic rate of increase (r), carrying capacity (K), and body size seemed to have any relationship with the probability of establishment of insect invaders. However, because this relationship was not very strong, and the analysis was done at the ordinal level, Lawton and Brown concluded that it lacks predictive value.

Below, we analyse data on insect invasions in the United States. We ask: (i) Are introduced species non-randomly distributed among families? (ii) If the answer to question (i) is yes, which families have more introduced species than would be expected by chance alone? (iii) Is this pattern real and not just a statistical artefact?; (iv) If the answer to questions (ii) and (iii) is yes, what generates this pattern? Based on the results of this analysis, we argue that the opportunity for colonisation is a crucial factor determining which introduced insects arrive and survive.

2. MATERIALS AND METHODS

2.1 The Database

We compiled data on the numbers of non-indigenous insect species in each family from Kim and Wheeler (1991). Kim and Wheeler list all known non-indigenous insect species for the United States until the date of publication of their report. They consider a species to be non-indigenous if it is resident or probably resident in the U.S. and was not native to the U.S. prior to European exploration and settlement.

We then calculated the source pool of species available to colonise the U.S. — the total number of species in each family that were not originally native to the U.S. and were thus originally available for invasion — as

$$n_i = w_i - (us_i - x_i) \qquad \text{(EQUATION 2.1.1)}$$

where n_i is the number of source species in family i, w_i is the total number of species in family i in the world, us_i is the total number of species in the US in family i (including non-indigenous species), and x_i is the number of exotic species of family i in the US. The quantity $us_i - x_i$ is the number of native species in family i in the US; we subtracted this quantity from the total number of species in each family because these species are obviously not available to invade the US. We obtained most of our w_i values from Parker (1982), all our us_i values from Arnett (1985), and all our x_i values from Kim and Wheeler (1991).

2.2 Statistical methods

To test whether there is overall selectivity in our data (i.e., whether the non-indigenous species are non-randomly distributed among families) we used a log-likelihood ratio test (G-test). Usually, one can compare the G-statistic calculated in the test with the χ^2 critical value obtained from tables. However, in order to do this it is necessary that the smallest expected value in the contingency table be no lower than 5 (Sokal and Rohlf 1995). In our case, because we have to fill 949 cells (families) with 1863 invasive species, surely *most* of our expected values will be lower than 5. To circumvent this problem it is possible to use randomizations to generate an expected distribution of the G-statistic (Manly 1997). Thus, we first distribute species uniform-randomly among families; for that randomly generated contingency table, we calculate the G-statistic; we repeat this procedure 10,000 times. We then compare the G-statistic calculated for the real data with the distribution of the simulated G-statistics. If the G-statistic calculated for our data is an extreme value of the distribution of the simulated G-statistics, we can reject the null hypothesis of random distribution of species among families in our data.

Having answered the question of whether there was overall selectivity in our data, we asked which families have more introduced species than would be expected by chance. To answer this question we used a one-tailed binomial test. The binomial test calculates the binomial probability $P(x)$ that a random sample of size n drawn from a binomial population will contain x elements in one of the categories (each element with probability p) and $n - x$ elements in the other category (each with probability $q = 1 - p$) (Zar 1996):

$$P(x) = \binom{n}{x} p^x q^{n-x} \qquad \text{(EQUATION 2.2.1).}$$

Thus, for a family i, we calculated the probability $P_i(x_i)$ that x_i species would be introduced out of a total n_i source species (i.e. the number of species that were "available" to invade the US in the world pool; see equation 17.1). Finally, we calculated p as the total number of non-indigenous insects in the U.S. divided by the total insect source species in the world, i.e.,

$$p = \sum_{i=1} x_i \Big/ \sum_{i=1} n_i = 1{,}863/788{,}681 = 0.0024 \qquad \text{(EQUATION 2.2.2).}$$

Equation 17.2 gives the probability that exactly x elements will be in one of the categories. However, we are interested in the probability of a particular case or more extreme cases, i.e., the tail probabilities; whether we take the upper tail, lower tail or two-tail probabilities will depend on our null hypothesis (H_0). Since we want to test whether the occurrence of x elements in one of the categories with probability p in a random sample of size n is too high to be considered random, we clearly want an upper, one-tailed test. The binomial upper-tail probability for family i is given by

$$p_i(x_i) = \sum_{k=x_i}^{n_i} \left[\binom{n_i}{k} p^k q^{n_i - k} \right] \qquad \text{(EQUATION 2.2.3).}$$

There is always some probability that we can get a significant result (a binomial tail probability lower than the critical value) by chance alone, and, as we are performing multiple tests, we may expect some such artifactual results. One possibility for dealing with this problem is to correct the critical value against which each binomial probability is compared; we used one such test, the sequential Bonferroni correction (Rice 1989). This correction works as follows: we order the binomial probabilities from highest (least significant) to lowest (most significant); we then divide the critical value (e.g., $\alpha = 0.05$) by its rank. For example, since in our test we were interested only in which families have "too many" non-indigenous species, we included only those families with at least one non-indigenous species — a total of 170. Thus, the critical value corresponding to our lowest (most significant) tail probability was divided by 170, the second lowest by 169, and so forth.

6. Taxonomic Selectivity in Introduced Insects

Table 1. Insect families with binomial probabilities lower than their corresponding critical value. Families are ordered by their corresponding binomial probability. n_i is the number of world source species in family i; x_i is the number of invasive species in the U.S. in the family; $P_i(x_i)$ is the binomial probability for the family; the critical value is calculated as 0.05 divided by the count for the family (see text for a more detailed explanation).

Order	Family	n_i	x_i	$P_i(x_i)$	Count	Critical value
Coleoptera	Staphylinidae	27041	228	0.0000000000	170	0.0002941176
Diptera	Cecidomyiidae	2975	34	0.0000000000	169	0.0002958580
Diptera	Oestridae	31	7	0.0000000000	168	0.0002976190
Hymenoptera	Aphelinidae	4044	61	0.0000000000	167	0.0002994012
Hymenoptera	Encyrtidae	2391	62	0.0000000000	166	0.0003012048
Hymenoptera	Eulophidae	2537	44	0.0000000000	165	0.0003030303
Hymenoptera	Pteromalidae	2446	41	0.0000000000	164	0.0003048780
Hymenoptera	Tenthredinidae	2307	38	0.0000000000	163	0.0003067485
Lepidoptera	Tortricidae	3025	78	0.0000000000	162	0.0003086420
Thysanoptera	Thripidae	1279	43	0.0000000000	161	0.0003105590
Hymenoptera	Tetracampidae	35	6	0.0000000003	160	0.0003125000
Hymenoptera	Eurytomidae	871	15	0.0000000063	159	0.0003144654
Hymenoptera	Diprionidae	65	6	0.0000000140	158	0.0003164557
Coleoptera	Anobiidae	158	7	0.0000001424	157	0.0003184713
Hymenoptera	Aphidiidae	698	12	0.0000002040	156	0.0003205128
Orthoptera	Gryllotalpidae	63	5	0.0000004984	155	0.0003225806
Homoptera	Aphididae	2169	20	0.0000006071	154	0.0003246753
Coleoptera	Dermestidae	732	11	0.0000023576	153	0.0003267974
Siphonaptera	Pulicidae	163	6	0.0000032863	152	0.0003289474
Coleoptera	Coccinellidae	4126	25	0.0000392087	151	0.0003311258
Homoptera	Adelgidae	31	3	0.0000590869	150	0.0003333333
Diptera	Tephritidae	3743	23	0.0000634001	149	0.0003355705
Thysanoptera	Aeolothripidae	178	5	0.0000793799	148	0.0003378378
Coleoptera	Bruchidae	1479	13	0.0000847723	147	0.0003401361
Heteroptera	Anthocoridae	422	7	0.0000867729	146	0.0003424658
Hymenoptera	Cynipidae	572	8	0.0000901753	145	0.0003448276

The sequential Bonferroni correction is designed to account for the possibility of finding some binomial tests significant by chance alone. To test how well this correction was working, we used a simulation. For this simulation, we uniform-randomly selected 1863 "invasive" species from the 788,681 source species, recording the families of the selected species. We then calculated the binomial probabilities for each of the 949 families, and ordered them from lowest to highest. We repeated this procedure 100 times. We then calculated the mean for each category (the mean of the lowest binomial probabilities in each of the 100 iterations, the mean of the second lowest probabilities, and so on). We compared these means with the rank-ordered binomial probabilities we obtained for the real data. Thus, were the Bonferroni correction working properly, none of our simulated binomial probabilities should be lower than its corresponding critical value, and they should be higher than the significant binomial probabilities calculated from the real data.

The statistical tests described above were performed using computer programs written in C. The programs and the data are available from the first author upon request.

3. IS THERE TAXONOMIC SELECTIVITY IN OUR DATA?

By looking at the list of non-indigenous species in each family (Appendix), we find that all the non-indigenous species in the U.S. occur in just 170 of the 949 families. This fact suggests that the non-indigenous species may not be randomly distributed among families. In the log-likelihood ratio test, the G-value for our data is $G = 1705.04$. This value exceeds all 10,000 simulated G-values (which range between 458.68 and 709.11, with mean 563.81) (Fig. 1). Thus, we can reject at $p < 0.0001$ the null hypothesis of random distribution of introduced species among families. This is, of course, far from surprising. We know that species are not equal and that phylogeny matters (Faith 1992). Why would we then expect non-indigenous species to be randomly distributed among families? — such a result would have been surprising.

As the introduced species are not randomly distributed among families, we can go further and ask which families have more non-indigenous representatives than would be expected by chance. Using the upper-tail binomial test we see that 26 families have binomial probabilities lower than their corresponding critical values (Table 1).

4. IS THE PATTERN REAL?

Although our corrected binomial test tells us that 26 families have more non-indigenous species than we would expect, we want to be sure that this pattern is not a statistical artifact — in other words, that the Bonferroni correction adequately accounts for the fact that we performed multiple tests. For this test we used the simulations described above to calculate the binomial probabilities for randomly generated data. We find that none of the binomial probabilities calculated in our simulations is lower than any of the significant binomial probabilities calculated for the real data (Fig. 2). This means that the pattern we found is not a statistical artifact.

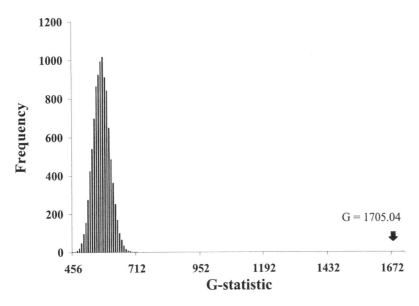

Figure 1. Distribution of the randomly generated *G*-statistics. The arrow shows, for comparison, the value of the *G*-statistic calculated for the real data.

5. WHAT IS PRODUCING THE PATTERN?

As the pattern is real, we can ask a more interesting question: what generates it? We must look at the particular families at the top of our list — those that show much higher numbers of non-indigenous species than expected (Table 1).

As we discussed in the introduction, several factors can determine which species will establish in a new location: the opportunity for colonisation, the characteristics (both biotic and abiotic) of the colonised habitat, and the biological characteristics of invaders. Below we analyze the only one of these factors that can be studied with the available data: the opportunity for colonisation. In particular, we ask whether the way species are transported can explain the pattern we observed in our data. Several families in our list are used in biological control, either as parasitoids (most Hymenoptera) or predators (Coccinellidae and Anthocoridae). Those families whose life history is linked to soil form a second group, and these may have arrived in soil brought from other continents, mainly from Europe and South America until the late nineteenth century. A third group comprises species that live on plants, mainly crops and other plants of economic importance. Several families fall in none of these groups; some are ectoparasites of vertebrates, scavengers, or predators of insects.

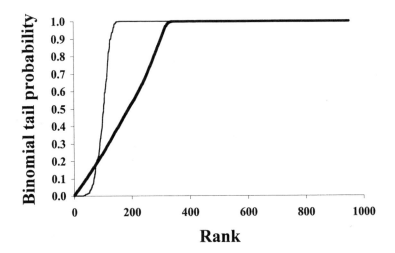

Figure 2. Rank-ordered binomial probabilities. Thick line: mean binomial probabilities calculated for the simulated data; thin line: binomial probabilities calculated for the real data (see the text for a more detailed explanation).

5.1 Biocontrol agents

Of the families listed in Table 1, only two were introduced to the U.S. as predators: the Coccinellidae and the Anthocoridae. Coccinellids are one of the first groups of insects used for biological control (Caltagirone and Doutt 1989), and they are the main biological agents introduced for the control of several homopteran plant pests, such as whiteflies (Aleyrodidae), scale insects (Coccidae), aphids (Aphididae), psyllids (Psyllidae), and several mites (Acari) (Obrycki and Kring 1998). Besides the coccinellids, a few species the Anthocoridae (Heteroptera) were also introduced as biocontrol agents of plant pests, such as the pear psylla (Clausen 1978). Of seven introductions of anthocorids into America north of Mexico, only two appear to have established (Lattin 1999). Most non-indigenous species in this last family were, however, apparently accidentally introduced (six of the seven species listed by Kim and Wheeler [1991]). To our knowledge, no predaceous insect families other than Coccinellidae and Anthocoridae have been introduced for biological control.

Five species of tephritid flies of the genus *Urophora* are listed in the North American Non-Indigenous Arthropod Database (NANIAD; Kim 1991) as having been introduced for the biological control of the knapweeds *Centaurea maculosa* and *C. diffusa*. Yet many other introduced tephritid flies are important pests of cultivated plants.

Two species of cecidomyiid flies were introduced for biological control of plants. But, again, the rest of the non-indigenous species in this family were accidentally introduced, probably with soil or introduced plants of economic importance (see below).

Of the ten Hymenoptera families listed in Table 1, seven are used as biological control agents in the United States. However, not all of the introduced species in those families are used for that purpose: the proportion of observed species mentioned by Clausen (1978; regrettably, the only comprehensive compendium of biological control agents introduced into the U.S. and world-wide) as having been purposefully introduced for biological control ranges between 0.25 and 0.67 (Table 2). Although Clausen's report is obviously outdated, it is likely that several of the species introduced in those families were actually accidental introductions. For example, Sailer (1983) mentions that, of the 232 *beneficial* Hymenoptera introduced in 1982, 35% arrived accidentally. So the numbers given by Clausen are probably not so far from reality. Frank et al. (1997) give two examples of parasitoids that arrived in Florida accompanying their hosts, which were also accidentally introduced: *Trichospilus distraeae* (Eulophidae), a wasp that parasitizes the introduced geometrid moth *Epimecis detexta*, and *Arrhenophagus albitibiae* (Encyrtidae), a parasite of scale insects. Although only the latter species has also been purposefully introduced for biological control (Kim 1991), these examples demonstrate the possibility of accidental introduction of parasitoids. On the other hand, while some diprionid parasitoids are used as biological control agents, others feed on plants and are considered pests (see below). Finally, Howarth (1991) argues that many of the apparently accidental introductions of biocontrol agents in Hawaii were actually unreported purposeful introductions. If Howarth's claim is true and typical beyond Hawaii, it might well be possible that some of Sailer's 35% accidental introductions of biological control agents were indeed purposeful unreported introductions.

If biological control were important in determining which families will be "selected" to be introduced (i.e., which will be at the top of our list), we would expect those families to have proportionally more species introduced for biological control than those towards the bottom of the list. If we take the binomial tail probabilities of section 17.3 as a selectivity index (lower binomial probability would mean higher selectivity), then we would expect a negative correlation between this index and the proportion of species in a family introduced for biocontrol (we used the proportion and not the actual number because this number will of course depend on family size). Since many hymenopteran families are used as biological control agents, we can use the Hymenoptera as a model group to test our hypothesis. In Fig. 3 we have plotted the proportion of species that is known to have been introduced for biological control vs. the binomial tail probabilities. Far from a nice negative correlation, our r^2 is 0.0449, and the relationship is obviously non-linear. However, if we look at Fig. 3, we do see that for very low values of binomial probabilities (i.e., those that are statistically significant) the proportion of species introduced for biological control seems to increase. If we use a statistical test to compare those families that have a statistically significant binomial probability with those that do not, we see that the two groups have significant differences in the proportion of species introduced for biological control (non-parametric two-sample Wilcoxon test, $p = 0.0064$).

5.2   Species associated with soil

Many, probably most, of the early insect introductions to the United States were of species associated with soil brought as ship ballast (Lindroth 1957). Early ship traffic came mainly from European ports — which explains the predominance of European species in early invasions (Sailer 1983). Later, in the late nineteenth century, with the increased commerce with South American countries, many new introductions came from that continent.

Mole crickets (Orthoptera, Gryllotalpidae) were introduced from South America, most likely in soil brought as ship ballast (Sailer 1983). Staphylinid beetles live under dead trees and are usually associated with soils (Borror et al. 1989), which makes them also candidates to have been introduced with ship ballast.

Finally, several cecidomyiid flies feed on grasses imported to the U.S. It is possible that these species came in soil brought with those grasses (Kim 1991).

5.3   Crop pests and other herbivores

Many herbivorous insect species have been introduced with their host plants. The Tenthredinidae are leaf-mining sawflies that feed on tree species such as the birch and hazelnut (Clausen 1978). All introductions in this family were accidental. Four tenthredinid sawflies are considered major pests: *Caliroa cerasi*, *Heterarthrus nemoratus*, *Hoplocampa brevis*, and

Pristiphora erichsonii; 16 are considered minor pests and the remaining 18 have no apparent economic impact (Kim 1991).

The Thripidae and Aelothripidae (Thysanoptera) are, in many cases, associated with economically important plant species. For example, the onion thrips (*Thrips tabaci*) is a pest of many cultivated plant species (Clausen 1978). Most introduced thrips found in Florida (probably the most important port of entry for thrips) are associated with cultivated plants (Frank et al. 1997).

Eurytomidae (Hymenoptera) present a wide variety of biologies. Most species are either endophytic phytophages or parasitoids of phytophagous insects (Gauld and Bolton 1988, Whitfield 1998). Most non-indigenous eurytomids were probably accidentally introduced. Of the 15 species listed by Kim and Wheeler (1991) as having been introduced into the U.S., none seems to have been purposefully introduced (Kim 1991); yet the ROBO Database (B.C.D.C. 1999) lists four eurytomids introduced for biocontrol between 1981-1983, although those introductions probably failed.

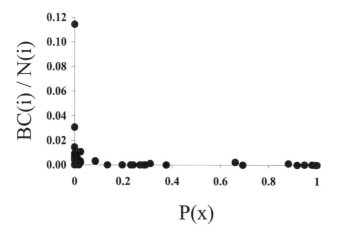

Figure 3. Proportion of species in each family of Hymenoptera known to have been introduced for biological control vs. the binomial tail probabilities (see section 17.3). Only families with at least one invasive species are included in the graph.

Most tortricid moths (if not all) are, once more, associated with introduced plants, mostly cultivated, such as apple, several berries, plum, currant, etc. (Kim 1991). For example, the omnivorous leaf-tier (*Cnephasia longana*) attacks many different species of crops, weeds, and native plants (Clausen 1978). Yet some species are associated with highly noxious non-indigenous plant species, such as Scotch broom.

Most non-indigenous aphids were introduced to the U.S. with their host plants. Virtually all these species are either major or minor pests of cultivated plants (Kim 1991).

Fruit flies (Tephritidae) are an important group of fruit and vegetable pests. Twenty three species are listed by Kim and Wheeler (1991) as introduced into the U.S., of which around 15 are pests. Interestingly enough, several species in the same family (*Urophora*) were introduced for biological control of weeds.

Of the six non-indigenous diprionid sawflies (Hymenoptera), four are plant feeders and pests of trees, such as most species of spruce (*Picea*) and pine (Clausen 1978). We have no information on how these species arrived in the U.S., although they probably came to North America in European trees imported in the 1920s to Ottawa, where they were first recorded (Clausen 1978). The other two non-indigenous diprionids were purposefully introduced for biological control (see above).

Most introduced bruchid beetles are pests, feeding mainly on seeds, such as beans, peas and other leguminous crops (Clausen 1978). They are believed to have entered the U.S. in their host seeds. Some species feed on invasive plants such as *Cytisus scoparius* (Scotch broom) (Kim 1991).

Although, as we mentioned above, two cecidomyiid flies were introduced for biological control, most species were accidentally introduced with plants or soil. And, again, one species is listed by Kim (1991) as feeding on brooms of the genus *Cytisus* (most likely Scotch broom).

Aphids of the family Adelgidae were accidentally introduced from Europe (Clausen 1978), probably as hosts of imported European plant species. The three species listed by Kim and Wheeler (1991) are listed as major pests by Kim (1991).

Finally, the Cynipidae (Hymenoptera) are gall-forming wasps that were most likely accidentally introduced with cultivated plants.

5.4   Other families

The Oestridae (Diptera) are internal parasites of mammals, especially of livestock. Four species in the genus *Gasterophylus* are intestinal parasites of horses; two species of *Hypoderma* are parasites of cattle and caribou; and one species of *Oestrus* is a parasite of sheep (Borror et al. 1989). All seven species are believed to have been introduced with their hosts (Frank et al. 1997).

The fleas (Siphonaptera, Pulicidae) are ectoparasites of birds and mammals and, again, are believed to have arrived with their hosts (Borror et al. 1989). These introduced pulicid fleas are cosmopolitan, having also been introduced by human transport to all continents except Antarctica (Lewis 1995).

The dermestid beetles are mainly scavengers, feeding primarily on materials rich in protein. Most introduced species are pests of stored materials such as wool, silk and dry food. Some species, notably *Athrenus*

6. Taxonomic Selectivity in Introduced Insects

museorum, feed on dead animal tissues and are a serious threat to museum collections. All non-indigenous dermestids probably arrived in the U.S. in their food sources.

Table 2. Species with significantly higher numbers of invasive species than the random expectation. Letters between brackets indicate orders: C, Coleoptera; D, Diptera; H, Hymenoptera; He, Heteroptera; Ho, Homoptera; L, Lepidoptera; O, Orthoptera; S, Siphonaptera; T, Thysanoptera. Numbers between brackets in the biological control column indicate numbers of species intentionally introduced for biocontrol over the total of introduced species in the family, according to Clausen (1978).

Biological control	Soils	Introduced with cultivated plants — crop pests	Other
Cecidomyiidae (D) (2/23)	Staphylinidae (C)	Bruchidae (C)	Oestridae (D) (endoparasites of mammals)
Tephritidae (D) (5/23)	Cecidomyiidae (D)	Cecidomyiidae (D)	Pulicidae (S) (ectoparasites of mammals)
Aphidiidae (H) (3/12)	Gryllotalpidae (O)	Tephritidae (D)	Anobiidae (C) (mainly stored food feeders)
Aphelinidae (H) (37/61)		Aphididae (Ho)	Dermestidae (C) (scavengers; dead animal tissue feeders)
Diprionidae (H) (2/6)		Adelgidae (Ho)	
Encyrtidae (H) (35/62)		Cynipidae (H)	
Eulopidae (H) (14/44)		Diprionidae (H)	
Pteromalidae (H) (16/41)		Eurytomidae (H)	
Tetracampidae (H) (4/6)		Tenthredinidae (H)	
Anthocoridae (He) (1/7)		Tortricidae (L)	
Coccinellidae (C) (12/25)		Thripidae (T)	
		Aeolothripidae (T)	

All introduced anobiid beetles are considered pests, feeding on stored products, mainly food. Probably the best-known species is the drugstore beetle, *Stegobium paniceum*, which is known for eating "almost everything", from food to books (Borror et al. 1989).

6. DISCUSSION

Perhaps the most obvious result of our analysis is that there is a highly non-random distribution of invasive species among insect families. This result has important implications for the theme developed in this book: by favoring the spread of only species clustered in certain families we are contributing to the homogenization of the world's biota — the production of a "planet of weeds" (Quammen 1998).

A second important result of our analysis is that, for those families that have more introduced species than would be expected by chance, human transport seems to be an important "selectivity factor". All the families at the top of our list (Table 1) had substantial numbers of species introduced either purposefully for biological control or accidentally as hosts of introduced plants, soils, animals or food products. Others have suggested the role of human transport as an important factor determining the introduction of propagules of invasive species. For example, Chown et al. (1998) have found a strong relationship between the number of human occupants and the number of species introduced in islands of the southern ocean; they found this to be the case for plants, insects, and mammals, but not for land birds. Lockwood (1999) also found human transport to be the most important factor determining the introduction of non-indigenous birds at a global scale. Finally, Williamson (1996) argues that human introductions are a crucial factor determining propagule pressure, which ultimately determines the establishment of introduced species. Thus, human transport would increase the likelihood of establishment of an invasive species by increasing propagule pressure. The role of propagule pressure as a determinant of the establishment of invasive species has been formally analysed by Hooper and Roush (1993). They found that, for parasitoids introduced for biological control of lepidopteran pests, the proportion of parasitoid populations that established increased with the number of parasitoids per release, the total number released, and the number collected.

It could be argued that the role of human transport might also be important for those families that show no selectivity (i.e., that the number of human-introduced species per family does not differ from the random expectation). Because of the difficulty of gathering information on transport for all 170 families with at least one introduced species, we tested this hypothesis only for the Hymenoptera. Within this order, the proportion of species in a family that is used for biological control seems to be related to the overall selectivity operating on that particular family. It could be argued that biological control agents are not "true invasives," because they are purposefully introduced to control particular target pest species, and therefore they do not have an important impact on the native biota. However, biological control agents do indeed colonise native, non-target hosts.

Hawkins and Marino (1997) found that 16% of parasitoids introduced into North America have colonised native hosts. Furthermore, owing to the characteristics of the available database, they judged this figure to be an underestimate.

It is noteworthy that many (most) non-biological control insects were introduced accidentally on their hosts. In most cases, those hosts are economically important plants, vertebrates or food products, although there are some examples in which the host itself was accidentally introduced (e.g., three species that feed on Scotch broom). We know that introduced host species also show taxonomic selectivity, at least in the case of plants (Daehler 1998, Pysek 1998); we hypothesise that the taxonomic selectivity observed in our insect database could be driven, at least in part, by the taxonomic selectivity operating on their hosts.

We must consider another alternative explanation for our finding: it is possible that the patterns we found, even though they are statistically robust, simply reflect which data are gathered. In particular, species important to humans either as pests or as useful tools for biocontrol are noticed more than species that do not have such direct importance. Although we believe that this is scenario is quite possible, again, because of the state of the available database, this hypothesis is untestable.

An important drawback in our data (as well as in most available data on species introductions) is that they are adequate only for those species that arrived and survived, not those that arrived but disappeared. For insects, only for biological control are there substantial data on failed introductions. Furthermore, we have no systematic data to quantify the impact of the different survivors on the native ecosystems, although absence of standardized impact data plagues all research on non-indigenous species (Parker et al. 1999).

Another possibility that we did not analyze is whether the biology of the invaders determines their success. Leston (1957) suggested that high level taxa have inherent properties that predispose some of them to be better colonists than others. Daehler (1998) found that some traits were indeed common among plant families that are over-represented in terms of the number of invasive species they contain. He found that over-represented families include: (i) species belonging to freshwater aquatic plant families, which are more likely to be both weedy and natural area invaders than species from other plant families; (ii) other weedy species that posses several of the life history characteristics of "ideal weeds" proposed by Baker (1965), such as flowering at an early age, high seed production, facultative self-pollination, and rapid vegetative growth; (iii) other natural area invaders that are grasses, nitrogen fixers, climbers, and clonal trees. Regrettably, as we mentioned in the introduction, other attempts to find those biological characteristics have been unsuccessful.

Nevertheless, we believe that the biological characteristics of insect invaders are potentially important determinants of their success after their introduction; however, the current development of the database precludes any generalisation. Furthermore, the very fact that we are working at higher taxonomic levels makes any attempt at finding such generalisations likely to fail. However, where the search for biological characteristics is done at lower taxonomic levels, and particularly when comparing closely related species within a taxonomic group, more useful generalisations may arise. Rejmánek's (1996) analysis of several reproductive morphological traits in closely related *Pinus* species is a good example. We believe that this approach — the detailed study of particular, closely related groups — is more promising than the approach taken by us and by several other authors in the present volume.

Ten years ago, Simberloff (1989) concluded that "it is depressing to be unable to draw striking generalisations about introduced insects but it would serve no worthwhile purpose to generalize prematurely". We believe that, in these respects, our degree of knowledge about insect invasions — and of invasions in general — has not changed significantly. The only possible generalisation, if any, is that human transport is an important factor in promoting the biotic homogenization of our changing world. We view this conclusion as neither new nor striking. However, until there are more comprehensive data on introductions, including those that fail, useful analysis of the roles of the biotic and abiotic nature of the location of introductions and the biology of introduced species are impossible.

ACKNOWLEDGMENTS

We appreciate the editors' invitation to contribute to this volume. We especially thank Dr. Kenneth Lakin for providing a copy of Kim and Wheeler's report. We also thank Florencia Fernández Campón, Tadashi Fukami, Leah Gibbons and two anonymous referees for comments on the manuscript, and Dr. Louis Gross for giving us access to the computer facilities of The Institute of Environmental Modeling at the University of Tennessee. D.P.V. thanks the Fulbright Commission and the Institute of International Education of the United States for financial support.

Appendix. Insect families with at least one species introduced in the United States (according to Kim and Wheeler 1991).

Order	Family	w_i	us_i	n_i	x_i	$P_i(x_i)$
Ephemeroptera	Baetidae	520	147	374	1	0.593
Orthoptera	Tettigoniidae	375	243	134	2	0.042
Orthoptera	Gryllidae	800	96	705	1	0.816

6. Taxonomic Selectivity in Introduced Insects 119

Order	Family	w_i	us_i	n_i	x_i	$P_i(x_i)$
Orthoptera	Gryllotalpidae	65	7	63	5	5×10^{-7}
Mantodea	Mantidae	1500	14	1490	4	0.480
Blattodea	Blattidae	525	8	522	5	0.009
Blattodea	Blattellidae	1740	15	1728	3	0.783
Blattodea	Blaberidae	1020	9	1015	4	0.229
Dermaptera	Forficulidae	30	4	28	2	0.002
Dermaptera	Anisolabididae	300	9	294	3	0.035
Dermaptera	Labiidae	384	9	377	2	0.229
Isoptera	Kalotermitidae	293	18	276	1	0.485
Isoptera	Rhinotermitidae	159	9	151	1	0.304
Coleoptera	Anobiidae	450	299	158	7	1×10^{-7}
Coleoptera	Anthribidae	2600	79	2523	2	0.984
Coleoptera	Bostrichidae	700	64	643	7	0.001
Coleoptera	Brentidae	2300	5	2296	1	0.996
Coleoptera	Bruchidae	1500	34	1479	13	8×10^{-5}
Coleoptera	Buprestidae	15000	675	14333	8	1.000
Coleoptera	Byrrhidae	300	72	229	1	0.423
Coleoptera	Cantharidae	5000	468	4534	2	1.000
Coleoptera	Carabidae	30000	2271	27778	49	0.990
Coleoptera	Cerambycidae	35000	956	34051	7	1.000
Coleoptera	Chrysomelidae	35000	1481	33575	56	0.998
Coleoptera	Cleridae	4000	266	3737	3	0.994
Coleoptera	Coccinellidae	4500	399	4126	25	4×10^{-5}
Coleoptera	Colydiidae	1300	87	1214	1	0.946
Coleoptera	Curculionidae	50000	2614	47539	153	0.188
Coleoptera	Dermestidae	850	129	732	11	2×10^{-6}
Coleoptera	Derodontidae	19	7	13	1	0.031
Coleoptera	Elateridae	9000	885	8123	8	0.999
Coleoptera	Eucnemidae	1200	71	1130	1	0.934
Coleoptera	Histeridae	3000	499	2503	2	0.983
Coleoptera	Hydrophilidae	2000	284	1721	5	0.397
Coleoptera	Lampyridae	2000	124	1881	5	0.471
Coleoptera	Lathridiidae	500	120	382	2	0.234
Coleoptera	Merylidae	5000	520	4481	1	1.000
Coleoptera	Nitidulidae	3000	183	2823	6	0.670
Coleoptera	Oedemeridac	1000	86	917	3	0.378
Coleoptera	Pselaphidae	5000	654	4349	3	0.998
Coleoptera	Scarabaeidae	25000	1395	23641	36	0.999
Coleoptera	Silphidae	175	42	134	1	0.275
Coleoptera	Staphylinidae	30000	3187	27041	228	0.000
Coleoptera	Tenebrionidae	18000	1008	16995	3	1.000
Coleoptera	Trogossitidae	600	64	538	2	0.370

Order	Family	w_i	us_i	n_i	x_i	$P_i(x_i)$
Siphonaptera	Ceratophyllidae	760	125	639	4	0.070
Siphonaptera	Pulicidae	190	33	163	6	3×10^{-6}
Diptera	Agromyzidae	1800	188	1614	2	0.899
Diptera	Anthomyidae	1000	148	861	9	3×10^{-4}
Diptera	Braulidae	4	1	4	1	0.010
Diptera	Calliphoridae	3000	78	2928	6	0.703
Diptera	Cecidomyiidae	4000	1059	2975	34	0.000
Diptera	Ceratopogonidae	1200	463	739	2	0.530
Diptera	Chamaemyiidae	100	36	66	2	0.011
Diptera	Chloropidae	1000	273	728	1	0.826
Diptera	Cryptochetidae	20	1	20	1	0.047
Diptera	Culicidae	3000	150	2853	3	0.967
Diptera	Drosophilidae	1500	117	1384	1	0.964
Diptera	Ephydridae	1000	426	575	1	0.749
Diptera	Hippoboscidae	330	28	305	3	0.038
Diptera	Lauxaniidae	1200	135	1066	1	0.923
Diptera	Lonchaeidae	500	38	463	1	0.671
Diptera	Muscidae	3000	622	2388	10	0.066
Diptera	Oestridae	65	41	31	7	0.000
Diptera	Opomyzidae	50	13	38	1	0.087
Diptera	Otitidae	400	127	275	2	0.142
Diptera	Platystomatidae	1000	41	960	1	0.900
Diptera	Psilidae	200	34	167	1	0.331
Diptera	Rhagionidae	300	104	200	4	0.001
Diptera	Rhinophoridae	85	4	82	1	0.179
Diptera	Sepsidae	240	34	209	3	0.014
Diptera	Stratiomyidae	1400	254	1149	3	0.521
Diptera	Syrphidae	5000	874	4135	9	0.659
Diptera	Tachinidae	6000	1277	4743	20	0.013
Diptera	Tephritidae	4000	280	3743	23	6×10^{-5}
Diptera	Tipulidae	13000	1517	11484	1	1.000
Heteroptera	Alydidae	250	29	223	2	0.101
Heteroptera	Anthocoridae	500	85	422	7	9×10^{-5}
Heteroptera	Cimicidae	75	14	63	2	0.010
Heteroptera	Coreidae	1800	120	1683	3	0.768
Heteroptera	Cynidae	300	7	296	3	0.035
Heteroptera	Lygaeidae	3000	288	2718	6	0.635
Heteroptera	Miridae	10000	1777	8259	36	0.001
Heteroptera	Nabidae	300	48	253	1	0.456
Heteroptera	Pentatomidae	5000	247	4760	7	0.937
Heteroptera	Reduviidae	5000	106	4895	1	1.000
Heteroptera	Rhopalidae	142	36	108	2	0.028

6. Taxonomic Selectivity in Introduced Insects

Order	Family	w_i	us_i	n_i	x_i	$P_i(x_i)$
Heteroptera	Tingidae	1800	157	1649	6	0.208
Homoptera	Adelgidae	50	22	31	3	6×10^{-5}
Homoptera	Aleyrodidae	1156	99	1059	2	0.722
Homoptera	Aphididae	3500	1351	2169	20	6×10^{-7}
Homoptera	Cicadellidae	20000	2507	17553	60	0.005
Homoptera	Delphacidae	1300	145	1156	1	0.938
Homoptera	Diaspididae	1500	194	1314	8	0.015
Homoptera	Phylloxeridae	60	29	32	1	0.074
Homoptera	Pseudococcidae	1100	280	823	3	0.317
Hymenoptera	Andrenidae	4000	1199	2802	1	0.999
Hymenoptera	Anthophoridae	4000	920	3082	2	0.995
Hymenoptera	Aphelinidae	4000	17	4044	61	0.000
Hymenoptera	Aphidiidae	800	114	698	12	2×10^{-7}
Hymenoptera	Apidae	1000	57	944	1	0.897
Hymenoptera	Argidae	500	59	442	1	0.654
Hymenoptera	Bethylidae	1000	196	805	1	0.855
Hymenoptera	Braconidae	40000	1937	38170	107	0.062
Hymenoptera	Cephidae	102	12	92	2	0.021
Hymenoptera	Ceraphronidae	150	48	103	1	0.219
Hymenoptera	Chalcididae	1400	102	1303	5	0.206
Hymenoptera	Chrysididae	3120	227	2897	4	0.916
Hymenoptera	Cimbicidae	140	12	129	1	0.267
Hymenoptera	Colletidae	3000	153	2848	1	0.999
Hymenoptera	Cynipidae	1200	636	572	8	9×10^{-5}
Hymenoptera	Diprionidae	100	41	65	6	1×10^{-8}
Hymenoptera	Encyrtidae	2800	471	2391	62	0.000
Hymenoptera	Eucoilidae	700	80	624	4	0.065
Hymenoptera	Eulophidae	3000	507	2537	44	0.000
Hymenoptera	Eupelmidae	700	95	610	5	0.017
Hymenoptera	Eurytomidae	1100	244	871	15	6×10^{-9}
Hymenoptera	Evaniidae	500	11	491	2	0.330
Hymenoptera	Figitidae	250	60	193	3	0.012
Hymenoptera	Formicidae	14000	696	13353	49	0.003
Hymenoptera	Ichneumonidae	15000	3322	11726	48	4×10^{-4}
Hymenoptera	Megachilidae	3000	682	2326	8	0.201
Hymenoptera	Mymaridae	1200	120	1088	8	0.005
Hymenoptera	Pamphiliidae	170	72	99	1	0.212
Hymenoptera	Platygastridae	1100	192	915	7	0.007
Hymenoptera	Pteromalidae	2800	395	2446	41	0.000
Hymenoptera	Scelionidae	3000	275	2727	2	0.989
Hymenoptera	Signiphoridae	75	21	55	1	0.124
Hymenoptera	Siricidae	90	19	74	3	0.001

Order	Family	w_i	us_i	n_i	x_i	$P_i(x_i)$
Hymenoptera	Sphecidae	7700	1139	6571	10	0.952
Hymenoptera	Tenthredinidae	3000	731	2307	38	0.000
Hymenoptera	Tetracampidae	35	6	35	6	3×10^{-10}
Hymenoptera	Tiphiidae	1500	225	1278	3	0.592
Hymenoptera	Torymidae	1000	175	831	6	0.016
Hymenoptera	Trichogrammatidae	440	43	401	4	0.017
Hymenoptera	Vespidae	800	415	391	6	0.000
Hymenoptera	Xiphydriidae	90	9	82	1	0.179
Trichoptera	Hydropsychidae	900	130	771	1	0.843
Trichoptera	Hydroptilidae	600	201	400	1	0.618
Lepidoptera	Agonoxenidae	4	1	4	1	0.010
Lepidoptera	Arctiidae	2000	264	1737	1	0.985
Lepidoptera	Argyresthiidae	100	52	49	1	0.111
Lepidoptera	Bombycidae	100	1	100	1	0.214
Lepidoptera	Choreutidae	400	29	372	1	0.591
Lepidoptera	Coleophoridae	500	169	334	3	0.047
Lepidoptera	Cossidae	1000	45	956	1	0.899
Lepidoptera	Gelechiidae	4000	630	3381	11	0.195
Lepidoptera	Geometridae	20000	1404	18604	8	1.000
Lepidoptera	Gracillariidae	1000	275	728	3	0.255
Lepidoptera	Hesperiidae	3000	290	2711	1	0.999
Lepidoptera	Lycaenidae	3000	136	2865	1	0.999
Lepidoptera	Lymantriidae	2500	35	2468	3	0.935
Lepidoptera	Lyonetiidae	300	122	179	1	0.350
Lepidoptera	Nepticulidae	400	82	319	1	0.535
Lepidoptera	Noctuidae	25000	2925	22085	10	1.000
Lepidoptera	Ochsenheimeriidae	23	1	23	1	0.054
Lepidoptera	Oecophoridae	4000	225	3788	13	0.131
Lepidoptera	Phaloniidae	500	110	391	1	0.609
Lepidoptera	Pieridae	2000	225	1776	1	0.986
Lepidoptera	Psychidae	6000	26	5976	2	1.000
Lepidoptera	Pyralidae	20000	1374	18649	23	1.000
Lepidoptera	Saturniidae	1000	69	932	1	0.893
Lepidoptera	Sesiidae	1000	115	888	3	0.359
Lepidoptera	Sphiingidae	850	124	727	1	0.826
Lepidoptera	Tineidae	3000	175	2828	3	0.965
Lepidoptera	Tortricidae	4000	1053	3025	78	0.000
Lepidoptera	Yponomeutidae	1000	32	977	9	0.001
Thysanoptera	Aeolothripidae	230	57	178	5	0.000
Thysanoptera	Phlaeothripidae	3000	348	2667	15	0.003
Thysanoptera	Thripidae	1500	264	1279	43	0.000

REFERENCES

Arnett, R. S. 1985. American Insects: A Handbook of the Insects of America North of Mexico. Van Nostrand Reinhold.

Baker, H. G. 1965. Characteristics and modes of origin of weeds. In: The Genetics of Colonizing Species (Baker, H. G., and Stebbins, C. L., eds.), Academic Press, Pp: 147-169.

B.C.D.C. 1999. Releases of Beneficial Organisms in the U.S. and Territories Database (ROBO). Biological Control Documentation Center, < http://www.ars-grin.gov/nigrp/robo.html>.

Borror, D. J., de Long, D. M., Triplehorn, C. A. 1989. An Introduction to the Study of Insects, sixth edition. Saunders College Publishing.

Caltagirone, L. E., and Doutt, R. L. 1989. The history of the vedalia beetle importation to California and its impact on the development of biological control. Annual Review of Entomology 34:1-16.

Chown, S. L., N.J.M. Gremmen, and K.J. Gaston. 1998. Ecological biogeography of southern ocean islands: species-area relationships, human impacts, and conservation. American Naturalist 152:562-575.

Clausen, C. P. (ed.) 1978. Introduced Parasites and Predators of Arthropod Pests and Weeds: A World Review. United States Department of Agriculture, Agriculture Handbook No. 480.

Daehler, C. C. 1998. The taxonomic distribution of invasive angiosperm plants: ecological insights and comparison to agricultural weeds. Biological Conservation 84:167-180.

Elton, C. S. 1958. The Ecology of Invasions by Animals and Plants. Methuen.

Faith, D. P. 1992. Conservation evaluation and phylogenetic diversity. Biological Conservation 61:1-10.

Frank, J. H., E.D. McCoy, H.G. Hall, G.F. O'Meara, and W.R. Tschinkel. 1997. Immigration and introduction of insects. In: Strangers in Paradise: Impact and Management of Non-indigenous Species in Florida (D. Simberloff, D.C. Schmitz, and T. Brown, eds.). Island Press, Pp: 75-99.

Gauld, I., and Bolton, B. (eds.). 1988. The Hymenoptera. British Museum (National History), Oxford University Press.

Hawkins, B. A., and P.C. Marino. 1997. The colonization of native phytophagous insects in North America by exotic parasitoids. Oecologia 112:566-571.

Hopper, K. R., and R.T. Roush. 1993. Mate finding, dispersal, number released, and the success of biological control introductions. Ecological Entomology 18:321-331.

Howarth, F. G. 1991. Environmental impacts of classical biological control. Annual Review of Entomology 36:485-509.

Kim, K. C. 1991. North American Non-Indigenous Arthropod Database. United States Department of Agriculture–Animal and Plant Health Inspection Service, < http://www.exoticforestpests.org/rppc>.

Kim, K. C., and A.G. Wheeler, Jr. 1991. Pathways and Consequences of the Introduction of Non-indigenous Insects and Arachnids in the United States. Report to the United States Congress Office of Technology Assessment (Contract # H3-6115.0).

Lattin, J. D. 1999. Bionomics of Anthocoridae. Annual Review of Entomology 44:207-231.

Lawton, J. H., and K.C. Brown. 1986. The population and community ecology of invading insects. Philosophical Transactions of the Royal Society of London B 314:607-617.

Leston, D. 1957. Spread potential and the colonisation of islands. Systematic Zoology 6:41-46.

Lewis, R. E. 1995. Fleas (Siphonaptera). In: Medical Insects and Arachnids (Lane, R. P., and Crosskey, R. W., eds.). Chapman & Hall, Pp: 529-575.

Lindroth, C. H. 1957. The Faunal Connections between Europe and North America. Almqvist & Wiksell.

Lockwood, J. L. 1999. Using taxonomy to predict success among introduced avifauna: relative importance of transport and establishment. Conservation Biology 13:560-567.

Lonsdale, W. M. 1999. Global patterns of plant invasions and the concept of invasibility. Ecology 80:1522-1536.

Manly, B. F. J. 1997. Randomization, Bootstrap and Monte Carlo Methods in Biology. Chapman & Hall.

Obrycki, J. J., and T.J. Kring. 1998. Predaceous Coccinellidae in biological control. Annual Review of Entomology 43:295-321.

Parker, I. M., D. Simberloff, W.M. Lonsdale, K. Goodell, M. Wonham, P.M. Kareiva, M., Williamson, B. Von Holle, P.B. Moyle, J.E. Byers, L. and Goldwasser. 1999. Impact: Toward a framework for understanding the ecological effects of invaders. Biological Invasions 1:

Parker, S. P. (ed.). 1982. Synopsis and Classification of Living Organisms. McGraw-Hill.

Pysek, P. 1998. Is there a taxonomic pattern to plant invasions? Oikos 82:282-294.

Quammen, D. 1998. Planet of weeds. Harper's 297:57-69.

Rice, W. R. 1989. Analyzing tables of statistical tests. Evolution 43:223-225.

Rejmánek, M. 1996. A theory of seed plant invasiveness: the first sketch. Biological Conservation 78:171-181.

Sailer, R. I. 1983. History of insect introductions. Pp. 15-38 In: Exotic Plant Pests and North American Agriculture (Graham, C., and Wilson, C., eds.). Academic Press.

Simberloff, D. 1986. Introduced insects: A biogeographic and systematic perspective. In: Ecology of Biological Invasions of North America and Hawaii (Mooney, H. A., and Drake, J. A., eds.), Springer-Verlag, Pp: 3-26.

Simberloff, D. 1989. Which insect introductions succeed and which fail? In: Biological Invasions: A Global Perspective (Drake, J. A., Mooney, H. A., di Castri, F., Groves, R. H., Kruger, F. J., Rejmánek, M., and Williamson, M., eds.). John Wiley & Sons, Pp: 61-72.

Sokal, R. R., and J.F. Rohlf. 1995. Biometry. W. H. Freeman and Company.

Whitfield, J. B. 1998. Phylogeny and evolution of host-parasitoid interactions in Hymenoptera. Annual Review of Entomology 43:129-151.

Williamson, M. 1996. Biological Invasions. Chapman & Hall.

Williamson, M. H., and A. Fitter. 1996. The characteristics of successful invaders. Biological Conservation 78:163-170.

Zar, J. H. 1996. Biostatistical Analysis. Prentice Hall.

Chapter 7

Are Unsuccessful Avian Invaders Rarer in Their Native Range Than Successful Invaders?

Thomas Brooks
Center for Applied Biodiversity Science, Conservation International, 2501 M Street, NW, Suite 200, Washington DC 20037, USA

1. INTRODUCTION

Biological invasions are a central component of the increasing taxonomic and geographic homogenization of the biosphere (Vitousek et. al., 1996), but our ability to predict such invasions remains in its infancy (Gilpin, 1990). A number of reviews (e.g., Ehrlich, 1986, 1989; Pimm, 1989; Lodge, 1993; Williamson, 1996) have listed the factors which may bias a species towards becoming a successful invader, including being abundant, widespread, and generalist. For example, Moulton and Pimm (1986) showed that large range size increased the chance of successful introduction into Hawai'i by passerine birds, and Scott and Panetta (1993) made a similar case for invasive plants in South Africa. Roy et. al. (1991) found a positive correlation between habitat generalism (measured by the number of climate zones within a species' native range) of biome grasses and number of continents invaded. Most recently, Goodwin et. al. (1999) found number of native biogeographic regions to be the best predictor of successful plant invasion into New Brunswick, Canada.

Such studies are rare, however: the number of quantitative tests of the effect of abundance, range size and niche breadth on invasion success remains tiny. This is largely, it seems, because data on these traits are scarce. One way in which such data can be made available is to invert the tests so as to consider the effects of species traits on introduction failure. Data on rarity do exist, for birds at least, thanks to the current urgency of conservation biology and especially to the research of BirdLife International. In this paper I therefore ask whether the probability of a species' introduction failing is increased by the species' being scarce, geographically restricted or ecologically specialist. The combinations of these traits comprise the "seven forms of rarity" (Rabinowitz, 1981; Rabinowitz et. al., 1986) and are generally not independent (Holt et. al., 1997; Gaston, 1996).

To set up these tests within a hypothesis-testing framework, a straightforward null hypothesis would be that the proportion of unsuccessfully introduced species that are rare is no different from that for successfully introduced species. Accepting this null hypothesis would imply that species traits are determined by interaction with the rest of community in which they are present. Thus, a species rare in one community could be common in another, and multiple introductions of a single species need not have similar outcomes (Moulton and Sanderson, 1997). A range of alternative hypotheses state that any combination of scarcity, range restriction or specialization affect (presumably detrimentally) a species' chances of successful introduction. Accepting one of these hypotheses would imply a more static characterization of a species: once rare, always rare. This would support the "all-or-none" patterns of species introduction success advocated by Simberloff and Boecklen (1991). It would also give hope to the search for characteristics useful in predicting potential invasive species.

2. METHODS

I used Long's (1981) exhaustive compilation of introduced birds of the world as my key source. Long (1981) lists all bird species introductions as definitely, probably or possibly successful or definitely, probably or possibly unsuccessful. I combined his 'probably or possibly' categories and categorized each species into the most successful category of introduction, regardless of the number of times introduced in each category. Thus, I listed Sulphur-crested Cockatoo *Cacatua galerita* as a successful introduction because it has been successfully introduced three times, even though another introduction of the species was only possibly successful and two more were unsuccessful (Long, 1981). In my initial analyses, I further combined the two successful and the two unsuccessful categories, although I retained these data to enable me to test the robustness of my results. I also noted the number if individual introduction attempts in each of the four categories (Appendices) to allow me to factor out the effects of differing numbers of introduction attempts—"propagule pressure" (Williamson, 1996).

I excluded four forms not identified to valid taxa by Long (1981) despite the fact that he listed them as unsuccessfully introduced: "'Golden' and 'Spectacled' Penguins", "Hummingbird spp.", and "'Mexican' or 'Butterfly' Bunting." Following Long's (1981) text (although inadvertently omitted from his tables) I did include Crimson-backed Tanager *Ramphocelus*

7. Are Unsuccessful Invaders Rare in Their Native Range?

dimidiatus, successfully introduced into Tahiti, and Sand Partridge *Ammoperdix heyi*, unsuccessfully introduced onto Cyprus; and also included Scarlet-rumped Tanager *Ramphocelus passerinii* listed but without text as unsuccessfully introduced into Tahiti (Long 1981).

No reliable abundance data exist for most bird species. Species that have "a high probability of extinction in the wild in the medium-term future," however, are listed in the summary Red List of threatened species (Collar et. al., 1994). Extinct species have populations of zero, and so it is only a modest assumption that species threatened with extinction must have small populations. I therefore categorized introduced species as scarce (Appendices) if they were listed by Collar et. al. (1994), either in one of the three threatened categories—critical (CR), endangered (EN), vulnerable (VU)—or as extinct (EX), extinct in the wild (EW), data deficient (DD), conservation dependent (CD), or near-threatened (NT).

Range size data for rare species have also been made available by BirdLife International, through Stattersfield et. al.'s (1998) list of all bird species with ranges smaller than 50,000 sq. km. I categorized introduced species as range restricted (noting the 'Endemic Bird Areas' or 'Secondary Areas' in which they are found) following Stattersfield et al. (1998) exactly (Appendices).

The number of habitats in which a species occurs must be a reasonable surrogate for its degree of ecological specialization. Sibley and Monroe (1990) give a habitat classification for all of the world's bird species, and so I categorized introduced species exactly following the habitats that they list (Appendices). None of these species were modified by Sibley and Monroe (1993).

I updated the taxonomy of Long (1981) in several cases. Brown Quail *Coturnix australis* and Swamp Quail *C. ypsilophora*, the former successfully introduced into New Zealand and Fiji from Australia and the latter probably unsuccessfully introduced into New Zealand from Tasmania (Long 1981) are lumped by Sibley and Monroe (1990). In contrast, Sibley and Monroe (1990) separate Canary-winged Parakeet *Brotogeris versicolorus* and Yellow-chevroned Parakeet *B. chiriri*, both of which are now well-established in Florida (Smith and Smith, 1993) and California (Garrett, 1993), and Laysan Finch *Telespiza cantans* and Nihoa Finch *T. ultima*, both of which have been the subject of re-introduction programs in Hawai'i (Long, 1981). Three further changes made by Sibley and Monroe (1990) compared to Long (1981) make no difference to my analyses: they split the *Alectoris graeca* complex, of which the Chukar *A. chukar* has been widely introduced; they lump the Edible-nest Swiftlet *Collocalia inexpectata*, introduced unsuccessfully into Hawai'i into the *C. fuciphaga* complex; and they split the Blue-hooded Euphonia *Euphonia musica* group of which Antillean

Euphonia *E. musica sensu stricto* has been unsuccessfully re-introduced onto Vieques Island from Puerto Rico.

In addition, Stattersfield et. al. (1998) split two species considered single species by Long (1981) and Sibley and Monroe (1990). They separate the threatened (Collar et. al., 1994) Gough Moorhen *G. comeri* from the extinct Tristan Moorhen *G. nesiotis*; the former was introduced onto Gough after the extinction of *G. nesiotis* (Long, 1981). They also split Mauritius Bulbul *Hypsipetes olivaceus* from Olivaceus Bulbul *H. borbonicus*; the latter was unsuccessfully introduced into the Chagos Archipelago (Long, 1981), and is considered threatened (Collar et. al., 1994).

3. RESULTS

In total, Long (1981) lists 259 bird species as successfully introduced (including probable and possible successes) and 174 species as unsuccessfully introduced. I classify each of these by rarity (scarcity, range restriction and ecological specialization). I then compared the proportions of rare species that were unsuccessfully introduced with corresponding proportions of successfully introduced species, using a Chi-square test with the proportion successfully introduced as the expected value. Classifications are provided in the appendices.

Of the species listed by Long (1981), 57 are listed by Collar et. al. (1994) and therefore considered to have small populations. Of these 31/259 (12%) were successfully introduced and 26/174 (15%) unsuccessfully introduced. These proportions are not significantly different ($\chi^2 = 1.35$, 1 d.f., P = 0.25). How robust is this result to our definition of threat? If we consider scarce only species listed as extinct, extinct in the wild or threatened (i.e., excluding species listed as CD or NT) by Collar et. al. (1994), we are left with 16/259 (6%) scarce species that were successfully introduced and 13/174 (7%) that were unsuccessfully introduced. Again, these proportions are not significantly different ($\chi^2 = 0.39$, 1 d.f., P = 0.60).

A total of 38 of the species listed by Long (1981) have ranges of less than 50,000 sq. km according to Stattersfield et. al. (1998) and are therefore considered to have restricted ranges. Of these 25/259 (10%) were successfully introduced and 13/174 (7%) unsuccessfully introduced. These proportions are not significantly different ($\chi^2 = 1.04$, 1 d.f., P = 0.30), although it is noteworthy that the proportion of successfully introduced species that had restricted ranges was marginally *higher* than that for unsuccessfully introduced species. I test the robustness of this result to the definition of range restriction by repeating the test considering only those species endemic to a single 'Endemic Bird Area' or 'Secondary Area'

7. Are Unsuccessful Invaders Rare in Their Native Range? 129

(Stattersfield et. al. 1998) to be range restricted. This leaves 20/259 (8%) successfully introduced species that had restricted ranges, and 11/174 (6%) unsuccessfully introduced species, which once more are not significantly different ($\chi^2 = 0.70$, 1 d.f., P = 0.45).

Following Sibley and Monroe's (1990) habitat classification, 56 of the species listed by Long (1981) are restricted to single habitats and therefore considered specialists. Of these, 24/259 (9%) were successfully introduced, and twice this proportion at 32/174 (18%) were unsuccessfully introduced. The latter proportion is significantly greater ($\chi^2 = 15.47$, 1 d.f., P < 0.001). Further, this result is robust to our definition of habitat specialization. To test this I considered species that Sibley and Monroe (1990) list as occurring in one or two habitats to be habitat specialists. Using this definition, the proportion of successfully introduced species that are habitat specialists rises to 64/257 (25%) but the proportion of unsuccessfully introduced species that are habitat specialists rises to 74/174 (42%). Again, this difference is highly significant ($\chi^2 = 27.27$, 1 d.f., P < 0.001).

How robust are these results to varying the degree of confidence with which we accept that an introduction has been successful? I test this, following Long's (1981) criteria, by using strict and then lax criteria for success. For the former, I consider 'probably' and 'possibly successful' introductions to have been unsuccessful (yielding 249 unsuccessfully introduced species), for the latter I consider "uncertain' and 'possibly unsuccessful' introductions as successful (yielding 159 unsuccessfully introduced species). Using strict criteria for success, the proportion of unsuccessfully introduced birds is no different from that for successful introductions for scarcity ($\chi^2 = 3.25$, 1 d.f., P = 0.06) or for range restriction ($\chi^2 = 0.04$, 1 d.f., P = 0.80) but is significantly greater for ecological specialization ($\chi^2 = 59.63$, 1 d.f., P < 0.001). Using lax criteria, similarly, the proportion of unsuccessfully introduced birds is no different from that for successful introductions for scarcity ($\chi^2 = 2.14$, 1 d.f., P = 0.15) or range restriction ($\chi^2 = 0.66$, 1 d.f., P = 0.45) but is significantly greater for ecological specialization ($\chi^2 = 9.99$, 1 d.f., P = 0.002). My results are robust to varying the criteria for introduction success.

To factor the effects of propagule pressure out of my results, I re-ran my analyses using total number of introduction attempts rather than individual species. In total, Long (1981) lists 947 individual successful introductions and 832 individual unsuccessful introductions. For scarcity, 85/947 (9%) and 95/832 (11%) respectively were listed by Collar et. al., (1994). For range restriction, 33/947 (3%) and 39/832 (5%) respectively have ranges of less than 50,000 sq. km (Stattersfield et. al., 1998). For ecological specialization, 59/947 (7%) and 102/832 (12%) respectively were restricted to single habitats (Sibley and Monroe, 1990). In each case, the difference between the

Table 1. Classification by seven forms of rarity (Rabinowitz et al. 1986) and commonness (i.e., species that are abundant, widespread, and generalist) of introduced bird species of the world (Long 1981). a) Successful introductions. b) Unsuccessful introductions. c) Comparison of the proportions of unsuccessful introductions which are rare (for each form of rarity) with those expected given the proportions of successful introductions which are rare, using a Chi-square test with Yates' correction where expected values are less than five.

a. Successful	Widespread		Restricted	
	Generalist	Specialist	Generalist	Specialist
Common	204/259 (79%)	16/259 (6%)	7/259 (3%)	1/259 (0%)
Rare	12/259 (5%)	2/259 (1%)	12/259 (4%)	5/259 (2%)

b. Unsuccessful	Widespread		Restricted	
	Generalist	Specialist	Generalist	Specialist
Common	121/174 (69%)	23/174 (13%)	3/174 (2%)	1/174 (1%)
Rare	14/174 (8%)	3/174 (2%)	5/174 (3%)	4/174 (2%)

c. Chi-square (1 d.f.)	Widespread		Restricted	
	Generalist	Specialist	Generalist	Specialist
Common	$\chi^2 = 7.55$ $P = 0.005$	$\chi^2 = 13.96$ $P < 0.001$	$\chi^2 = 0.82$ $P = 0.45$	$\chi^2 = 0.25$ $P = 0.70$
Rare	$\chi^2 = 4.71$ $P = 0.04$	$\chi^2 = 2.26$ $P = 0.10$	$\chi^2 = 1.87$ $P = 0.30$	$\chi^2 = 0.89$ $P = 0.45$

proportions of rare species successfully and unsuccessfully introduced was greater than when only individual species were counted ($\chi^2 = 5.86$, 1 d.f., P = 0.02; $\chi^2 = 3.57$, 1 d.f., P = 0.06; $\chi^2 = 41.27$, 1 d.f., P < 0.001; respectively). Ecological specialization remains by far the most important determinant on introduction success, however.

How do the different categories of rarity for bird introductions interact with each other? In Table 1a and 1b, respectively, I give the proportions of species successfully and unsuccessfully introduced for each of the seven possible combinations of rarity, and for commonness (i.e., abundant, widespread and ecologically generalized). I tested whether the proportionate distribution of species across these combinations for unsuccessful introductions differed from that for successful introductions, using a Chi-square test with Yates' correction (Spiegel, 1994) where χ^2 (corrected) = ((⁄o - e⁄ - 0.5)²/e) in cases where the expected value was less than five. The difference was indeed significant ($\chi^2 = 24.78$, 7 d.f., P = 0.001). Inspecting Table 1 clearly shows that not all of these combinations differed from each other, so I tested each combination individually, again using a Chi-square test with Yates' correction where necessary. I give these results in Table 1c. Significantly fewer unsuccessfully introduced birds were common compared to successfully introduced birds, and significantly more unsuccessfully introduced birds were ecologically specialized only. Marginally higher proportions of unsuccessfully introduced birds were also ecological specialized and threatened, and threatened only.

4. DISCUSSION

The fundamental importance of rarity within ecological communities is increasingly widely recognized (Gaston, 1994) and several excellent analysis have now been carried out building on the framework of Rabinowitz et. al. (1986), for example, for birds in Colombia (Kattan, 1992) and Brazil (Goerck, 1997). This is the first time that such an analysis has been carried out for introduced species. Perhaps somewhat counterintuitively, rarity caused by scarcity and by small range size appears to have no effect on introduction success for birds. Rarity caused by ecological specialization, however, has a strong effect, with introductions of ecologically specialized species far less likely to succeed than introductions of generalists. These results are robust to varying definitions of the three types of rarity, and of introduction success. They are also concordant with the literature: the only previous global study of the effect of rarity on invasion success showed that ecological generalists were better colonizers than specialists (Roy et. al., 1991) and Bazzaz (1986) argues that this should be the case on theoretical grounds.

My results should be interpreted with several caveats in mind. First, of course, the distribution of rarity among introduced birds in general (whether successfully or unsuccessfully introduced) is in no way typical of that for all birds. This is easily shown for all three types of rarity. For scarcity, the 11% of the world's birds that are threatened (Collar et. al., 1994) is significantly more than the 29/433 (7%) for introduced species (χ^2 = 8.45, 1 d.f., P = 0.004). For range restriction, the 27% of birds that have ranges of less than 50,000 sq. km (Stattersfield et al. 1998) is significantly more than 38/433 (9%) for introduced species (χ^2 = 74.43, 1 d.f., P < 0.001). For ecological specialization, the ~35% of birds that are restricted to single habitats (Sibley and Monroe 1990) is significantly more than the 55/433 (13%) for introduced species (χ^2 = 95.15, 1 d.f., P < 0.001). This general commonness of introduced birds does not hold for all families—Lockwood et. al. (in press) show that four families (Anatidae, Columbidae, Phasianidae and Psittacidae) are both selectively threatened and selectively introduced.

Nevertheless, people's initial selection of avian candidates for introduction is overall strongly biased towards common species.

A second important point is that unsuccessful introductions may well not be reported as often as successful ones, and so the absolute numbers of unsuccessfully introduced species given by Long (1981) may be smaller than in reality. This should not affect my results, though, as there is no reason to suspect that failed introductions of rare species would be reported with any different frequency than failed introductions of common species.

Another potential bias could be from introductions that post-date Long (1981). For example, Lever (1987) adds four successful introductions: Yellow-crowned Night-heron *Nyctanassa violacea*, Goshawk *Accipiter gentilis*, Griffon Vulture *Gyps fulvus* and Great Horned Owl *Bubo virginianus*. While these four species are common, current advocacy by some of translocation (Franklin and Steadman, 1991) and reintroduction (Beck et. al., 1994) as major conservation tools, despite the low success of such projects (Griffith et. al., 1989), may have added a few rare species. The California Condor *Gymnogyps californicus* reintroduction program is a particular example (Toone and Wallace, 1994), although its success is in no way assured (Pitelka, 1981). Overall, recent changes to Long's (1981) list must surely be few, though.

While Stattersfield et. al., (1998) give a direct classifications of small range size, how effective are my surrogate classifications of scarcity and of ecological specialization? One worry is that using the summary Red List (Collar et. al., 1994) conflates range size (criterion D2) with population scarcity. Indeed, five successfully introduced species (Little Spotted Kiwi *Apteryx owenii*, Gough Moorhen *Gallinula nesiotis*, Chatham Island Snipe *Coenocorypha pusilla*, Nihoa Finch *Telespiza ultima* and Laysan Finch *T. cantans*) are listed as threatened solely on account of their tiny ranges, whereas no unsuccessfully introduced species are. Excluding these from my calculations would cause a significantly higher proportion of unsuccessfully introduced to be threatened ($X^2 = 3.96$, 1 d.f., $P = 0.05$), although this result is marginal, especially given the taxonomic uncertainty regarding *Telespiza* (Olson and James, 1986). Using Sibley and Monroe's (1990) habitat classifications seems less of a worry: while these are undoubtedly subjective to some degree, there is no reason why this subjectivity should bias my results.

What of the impact of propagule pressure? A particular worry is that people may increase introduction intensity for common species either deliberately, because common species are perceived to have a high chance of introduction success (Green, 1997), or accidentally, because common species are more likely to be transported at random (Goodwin et. al., 1999). My results are similar whether or not propagule pressure (as measured by number of individual introduction attempts) is factored out, suggesting that this is not a major problem (although when propagule pressure is factored out, scarcity as measured by threat does marginally increase the chances of unsuccessful introduction). Other factors including number of individuals introduced and human effort invested (e.g., through supplemental feeding) influence propagule pressure (Williamson, 1996), but it seems unlikely that these will have an effect where total numbers of introduction attempts does not. This is definitely not to say that propagule pressure has no impact on introduction success—there is a strong relationship between numbers of individual birds introduced and introduction success in New Zealand (Veltman et. al., 1996; Duncan, 1997; Green, 1997). Rather, the effect of propagule pressure appears to be independent of the (non-)effects of rarity on introduction success.

This result sheds interesting light on the recent finding by Lockwood (1999) that taxonomy is a strong predictor of successful avian introduction (*contra* Williamson, 1996). Various explanations can be conceived for this. One possibility, suggested by Russell et. al. (1998) to explain taxonomic selectivity in species extinctions, could be possible taxonomic clumping of traits such as abundance, range size and niche breadth. Webb et. al. (this volume) show that such traits are phylogenetically independent to a remarkable degree, however. Further, if such traits were taxonomically clustered, we would expect there to be a negative relationship between the number of species successfully introduced and the number unsuccessfully introduced in each family (Lockwood, 1999). Species in families sharing traits that increase propensity for introduction success would generally succeed and vice versa. In fact a positive relationship exists, suggesting that human preferences may drive the selection of particular families for attempted introduction, with the success of individual species simply due to more persistent introduction activities, i.e., propagule pressure (Lockwood, 1999). My finding supports this conclusion.

In conclusion, this study shows that the role of rarity in avian introduction success varies depending on the form of rarity considered. Population scarcity and range restriction appear to have no effect on introduction success, supporting the view that whether or not species introductions succeed depends largely on the context of the host community (Moulton and Sanderson, 1997). In contrast, ecological specialization does appear to have a significant negative effect on introduction success. While there is certainly some evidence for rapid evolutionary response by birds introduced into novel habitats (Diamond et. al., 1989), it seems that ecological specialization is the least labile type of rarity. This may suggest that an "all or none" pattern of invasion success (Simberloff and Boecklen, 1991) exists when species are classified by ecological specialization. If this is the case, then such classification may be a valuable tool for warning which are the generalist species likely to spread in our increasingly homogenized world (Lövei 1997).

ACKNOWLEDGMENTS

Many thanks to M. L. McKinney, J. L. Lockwood and C. M. Wilder for their help with this study.

Appendix I. Successfully introduced bird species. Taxonomy follows Sibley & Monroe (1990), except where modified by Stattersfield et al. (1998) as noted in the text. Nomenclature and systematic order follow Monroe & Sibley (1993). Numbers of introductions follow Long (1981) with A = definite success, B = probable or possible success, C = uncertain or probable failure, and D = definite failure. Threat (Thr.) follows Collar et al. (1994) with EX = extinct, EW = extinct in the wild, CR = critical, EN = endangered, VU = vulnerable, CD = conservation dependent, and NT = near-threatened. "Endemic Bird Areas" (EBA) and secondary areas (s) are numbered following Stattersfield et al. (1998). Habitat follows Sibley & Monroe (1990) with * = specialized to one habitat only and ** = specialized to two habitats.

Species	A	B	C	D	Thr	EBA	Habitats
Ostrich *Struthio camelus*	1	-	-	1	-	-	Grassland, savanna, steppe, bush
Lesser Rhea *Rhea pennata*	1	1	-	-	NT	-	**Puna, grasslands
Southern Cassorary *Casuarius casuarius*	2	-	-	-	-	-	Humid forest, edge, especially near streams
Emu *Dromaius novaehollandiae*	1	-	-	2	-	-	Plains, scrub, open woodland, coastal heath, alpine pasture, semi-desert
Brown Kiwi *Apteryx australis*	1	-	-	-	VU	-	Forest, scrub, overgrown farmland
Little Spotted Kiwi *Apteryx owenii*	1	-	-	-	VU	-	*Forest
Chilean Tinamou *Northoprocta perdicaria*	1	-	-	1	-	60	**Grassland, wheatfields
Plain Chacalaca *Ortalis vetula*	1	-	-	1	-	-	Tall brush, thickets, scrub, second growth
Rufous-vented Chacalaca *Ortalis ruficauda*	-	1	-	-	-	-	Thorny brushlands, forest, scrub, second growth
Great Currasow *Crax rubra*	-	1	-	4	-	-	*Humid forest
Australian Brush-turkey *Alectura lathami*	-	2	-	3	-	-	Humid forest, scrubby creek margins, scrub

7. Are Unsuccessful Invaders Rare in Their Native Range?

Species							Habitat
Chukar *Alectoris chukar*	11	2	5	4	-	-	**Dry rocky and stony hillsides with sparse vegetation, grassy slopes
Barbary Partridge *Alectoris barbara*	2	1	1	7	-	-	Rocky and scrubby hillsides, ravines with bushes or junipers, mostly in desert areas
Red-legged Partridge *Alectoris rufa*	5	3	1	6	-	-	Open rocky or scrubby country, farmlands
Black Francolin *Francolinus francolinus*	3	4	-	4	-	-	Grasslands, scrubby and bushy areas, marshes
Chinese Francolin *Francolinus pintadeanus*	2	-	5	-	-	-	Scrub, clearings, dry forest
Grey Francolin *Francolinus pondicerianus*	8	-	1	1	-	-	Open arid country with scrub or grass, fields
Red-billed Francolin *Francolinus adspersus*	-	1	-	-	-	-	*Dry bush country along watercourses
Red-necked Spurfowl *Francolinus afer*	1	-	-	-	-	-	**Brushy and grassy areas, woody gorges
Erckel's Francolin *Francolinus erckelli*	1	-	1	-	-	-	Scrub, brush, open areas with scattered trees
Grey Partridge *Perdix perdix*	5	-	1	8	-	-	Farmland, undergrowth of woodland edge, steppes, meadows
Daurian Partridge *Perdix dauuricae*	2	-	-	1	-	-	Forested steppes, farmlands, rocky slopes
Madagascar Partridge *Margaroperdix madagascarensis*	1	-	-	2	-	-	Brush, grassland, weeedy farmland
Common Quail *Coturnix coturnix*	2	2	-	9	-	-	Fields, meadows, pastures, grassy slopes
Brown Quail *Coturnix ypsilophora*	1	1	2	-	-	-	Swampy grasslands, heavy pasture, swampy heaths
Blue-breasted Quail *Coturnix chinensis*	-	5	2	2	-	-	Grasslands, marshy areas, scrub, forest clearings, dry farmlands
Chinese Bamboo-Partridge *Bambusicola thoracica*	5	-	-	3	-	-	**Dry bush country, bamboo thickets
Red Jungle Fowl *Gallus gallus*	4	1	3	12	-	-	Forest undergrowth, second growth, scrub, grassland, farmlands
Kalij Pheasant *Lophura leucomelanos*	1	-	-	3	-	-	Forest undergrowth near streams, thick undergrowth, bamboo, cane, dense scrub
Silver Pheasant *Lophura nycthemera*	-	1	-	13	-	-	Forest, grassy slopes at forest edge, bamboo thickets
Swinhoe's Pheasant *Lophura swinhoii*	-	1	1	-	NT	149	*Primary or mature secondary hardwood forest
Crested Fireback *Lophura ignita*	-	1	-	1	-	-	*Forest
Cheer Pheasant *Catreus wallichii*	3	-	-	1	VU	128	Forest, scrub, meadows

Species							Habitat
Reeve's Pheasant *Syrmaticus reevesii*	-	1	5	1	VU	-	Tall grass and bushes of open woodlands with pine, cypress, Thuja or oak.
Ring-necked Pheasant *Phasianus colchicus*	17	-	6	18	-	-	Open country, farmland, scrubby wastes, open woodland, edge, grassy steppes, desert oases, riverine thickets, swamps
Golden Pheasant *Chrysolophus pictus*	-	1	-	7	NT	-	Bushy slopes, bamboo, woodland, terraced fields
Lady Amherst's Pheasant *Chrysolophus amherstiae*	1	-	-	3	NT	-	Wooded slopes, bamboo, thickets, dense bushes
Common Peafowl *Pavo cristatus*	2	6	-	6	-	-	Open forest, edge, second growth
Spruce Grouse *Dendragapus canadensis*	-	3	-	-	-	-	**Coniferous forest, dense undergrowth
Blue Grouse *Dendragapus obscurus*	-	1	-	2	-	-	*Coniferous forest
Willow Grouse *Lagopus lagopus*	2	4	-	13	-	-	Tundra, subalpine and subantarctic moors
Black Grouse *Tetrao tetrix*	1	3	-	16	-	-	Forest edge, moors, swampy heathland
Capercaillie *Tetrao urogallus*	6	2	-	16	-	-	*Mature coniferous forest
Hazel Grouse *Bonasa bonasia*	-	1	-	1	-	-	*Forest
Ruffed Grouse *Bonasa umbellus*	9	1	-	4	-	-	*Deciduous and mixed forest
Sharp-tailed Grouse *Tympanuchus phasianellus*	-	1	-	4	-	-	Grasslands, arid sagebrush, scrub forest, oak savanna
Greater Prairie-chicken *Tympanuchus cupido*	-	1	-	5	-	-	Prairie, forest edge, sandy grasslands
Wild Turkey *Melagris gallopavo*	12	-	-	11	-	-	Deciduous and mixed forest, open woodland, savanna
Helmeted Guineafowl *Numida meleagris*	7	19	5	10	-	-	Open country in virtually all habitats except desert and forest
Mountain Quail *Oreortyx pictus*	2	1	-	2	-	-	Brushy montane forest, edge, adjacent chaparral
Scaled Quail *Callipepla squamata*	2	-	-	1	-	-	**Desert grasslands, thorn scrub
California Quail *Callipepla californica*	13	-	-	9	-	-	Grassland, semi-desert srcub, bushy areas, chaparral, sagebrush
Gambel's Quail *Callipepla gambelii*	3	1	1	-	-	-	**Desert, thorn scrub
Northern Bobwhite *Colinus virginianus*	9	4	6	16	-	-	Brushy fields, grasslands, farmlands, thickets
Crested Bobwhite *Colinus cristatus*	2	-	-	-	-	-	Savannas, thickets, grasslands, forest edge, pastures
Black-bellied Whistling-Duck *Dendrocygna autumnalis*	1	-	1	1	-	-	Marshes, lagoons, stream borders
Ruddy Duck	2	-	-	-	-	-	Marshes, lakes, in migration also

7. Are Unsuccessful Invaders Rare in Their Native Range?

Species					IUCN	Pop	Habitat
Oxyura jamaicensis							streams, brackish marshes and coastal bays
Mute Swan *Cygnus olor*	6	-	-	1	-	-	Lakes, ponds, marshes, sluggish rivers
Black Swan *Cygnus atratus*	2	-	-	3	-	-	Lakes, bays, flooded pastures, swamps
Trumpeter Swan *Cygnus buccinator*	1	-	1	-	NT	-	Ponds, lakes, marshes
Graylag Goose *Anser anser*	4	-	-	3	-	-	Reedbeds, marshy swamps, estuaries, lakes
Nene *Branta sandwichensis*	1	-	-	-	VU	218	*Uplands, primarily sparsely vegetated lava flows
Canada Goose *Branta canadensis*	11	-	1	2	-	-	Marshes, meadows, lakes, rivers, tundra, in migration also brackish coastal areas and bays
Cape Barren Goose *Cereopsis novaehollandiae*	1	-	3	-	-	-	Pastures, lake margins, damp scrub
Egyptian Goose *Alopochen aegyptiacus*	2	-	-	4	-	-	Lakes, marshes, swamps, rivers
Paradise Shelduck *Tadorna variegata*	3	-	-	-	-	-	Tussock river flats, streams, lakes
Wood Duck *Aix sponsa*	3	-	-	2	-	-	Wooded swamps, streams, marshes, in migration also flooded fields and brackish coastal marshes
Mandarin *Aix galericulata*	2	-	1	2	NT	-	Wooded ponds, swamps, marshes, rocky streams
Gadwall *Anas strepera*	3	-	-	-	-	-	Marshes, lakes, in migration also rivers, flooded fields and brackish coastal areas
Mallard *Anas platyrhynchos*	6	-	4	1	-	-	Ponds, lakes, marshes, flooded fields, streams, farmlands, seacoasts, bays
Meller's Duck *Anas melleri*	1	-	-	1	NT	95,96	Lakes, ponds, marshes
Northern Pintail *Anas acuta*	2	-	-	-	-	-	Lakes, rivers, marshes, ponds, barrens, tundra
Redhead *Aythya americana*	1	-	-	-	-	-	Marshes, lakes, lagoons, rivers, especially with emergent vegetation, in migration also brackish coastal areas
Madagascar Buttonquail *Turnix nigricollis*	1	1	-	1	-	-	Grassland, savanna, brush, forest clearings
Laughing Kookaburra *Dacelo novaeguineae*	6	-	-	1	-	-	Open woodland, forest, farmlands, orchards, towns
Blue-streaked Lory *Eos reticulata*	-	1	-	-	NT	165	*Unknown, probably humid forest
Rainbow Lorikeet *Trichoglossus haematodus*	-	1	-	-	-	-	Humid and eucalyptus forest, woodland, swamps, scrub, heath, towns

Species							Habitat
Kuhl's Lorikeet *Vini kuhlii*	1	1	-	-	EN	211	*Coconut palms
Blue Lorikeet *Vini peruviana*	-	1	-	1	VU	213, 214,s 135	*Coconut palms
Musk Lorikeet *Glossopsitta concinna*	-	1	-	-	-	-	Eucalyptus forest, dry forest, riverine woodland
Galah *Eolophus roseicapillus*	1	-	-	2	-	-	**Open country with scattered trees, riverine woodland
Yellow-crested Cockatoo *Cacatua sulphurea*	-	2	-	-	EN	-	Forest edge, woodland, farmlands, coconut palms
Sulphur-crested Cockatoo *Cacatua galerita*	3	1	-	3	-	-	Forest, savanna, swamp, palm and eucylaptus forest, mangroves, farmlands
Salmon-crested Cockatoo *Cacatua moluccensis*	-	1	-	1	VU	170	**Forest, woodland
Tanimbar Cockatoo *Cacatua goffini*	-	1	-	-	NT	165	Forest, woodland, scattered trees
Long-billed Corella *Cacatua tenuirostris*	-	2	-	1	-	184	Open forest, woodland, riverine forest, farmlands
Great-billed Parrot *Tanygnathus megalorhynchos*	-	1	-	-	-	-	Forest, edge, second growth, farmlands
Eclectus Parrot *Eclectus roratus*	-	2	-	1	-	-	Forest, second growth, savanna, eucalyptus woodland
Red Shining-Parrot *Prosopeia tabuensis*	2	-	-	2	-	202	Forest, second growth, farmlands, towns
Crimson Rosella *Platycercus elegans*	2	-	-	1	-	-	Humid forest, woodland, scrub, towns
Eastern Rosella *Platycercus eximius*	2	-	-	-	-	-	Open forest, woodland, riverine forest, farmlands, towns
Red-rumped Parrot *Psephotus haematonotus*	-	1	-	-	-	-	Open woodland (usually near water), scrub, farmlands, towns
Budgerigar *Melopsittacus undulatus*	1	-	-	8	-	-	Grasslands, spinifex, mallee, mulga, riverine woodland, farmlands
Yellow-fronted Parakeet *Cyanoramphus auriceps*	1	-	-	-	-	-	**Forest, subalpine scrub
Meyer's Parrot *Poicephalus meyeri*	1	-	-	-	-	-	Savanna, riparian woodland, second growth, acacia scrub, farmlands
Grey-headed Lovebird *Agapornis canus*	1	3	-	3	-	-	Forest edge, brush, grasslands, farmlands
Fischer's Lovebird *Agapornis fischeri*	2	-	-	-	NT	108	Acacia grassland, savanna, farmlands
Yellow-collared Lovebird *Agapornis personatus*	2	-	-	1	-	-	**Acacia grasslands, savanna
Lilian's Lovebird *Agapornis lilianae*	-	1	-	1	-	-	*Mopane and acacia woodland, mostly in river valleys
Alexandrine Parakeet *Psittacula eupatria*	-	1	-	-	-	-	Forest, woodland, farmlands, mangroves

7. Are Unsuccessful Invaders Rare in Their Native Range?

Species							Habitat
Rose-winged Parakeet *Psittacula krameri*	6	9	-	5	-	-	Open woodland, savanna, farmlands
Red-breasted Parakeet *Psittacula alexandri*	1	1	-	1	-	-	Forest, second growth, mangroves
Orange-fronted Parakeet *Aratinga canicularis*	-	1	-	-	-	-	Deciduous forest, edge, open woodland, arid scrub, swamps, towns
Brown-throated Parakeet *Aratinga pertinax*	1	-	-	1	-	-	Arid scrub, semi-desert, mangrove, savanna, farmlands, woodland
Nanday Parakeet *Nandayus nenday*	-	3	-	-	-	-	Savanna, palm groves, farmlands
White-eared Conure *Pyrrhura leucotis*	-	1	-	-	-	-	*Humid forest
Monk Parakeet *Myiopsitta monachus*	2	-	-	2	-	-	Open forest, riverine woodland, acacia scrub, palm groves, orchards, farmland
Green-rumped Parrotlet *Forpus passerinus*	1	3	-	1	-	-	Savanna, open woodland, dry forest, edge, acacia scrub, mangroves, farmlands
Canary-winged Parakeet *Brotogeris versicolorus*	3	-	-	-	-	-	Open forest, edge,
Yellow-chevroned Parakeet *Brotogeris chiriri*	1	-	-	-	-	-	Open forest, edge, woodland, scrub, savanna
Orange-chinned Parakeet *Brotogeris jugularis*	-	1	-	2	-	-	Open woodland, second growth, forest edge, arid scrub
Hispaniolan Parrot *Amazona ventralis*	1	-	-	-	NT	28	**Forest, farmlands
White-fronted Amazon *Amazona albifrons*	-	1	-	-	-	-	Deciduous forest, open woodland, second growth, scrub, savanna, farmlands
Green-cheeked Amazon *Amazona viridigenalis*	-	2	-	-	EN	11	Forest, woodland, farmlands
Yellow-crowned Amazon *Amazona ochrocephala*	-	2	-	1	-	-	Forest, savanna, woodland, farmlands
Barn Owl *Tyto alba*	5	-	-	3	-	-	Open country, savanna, farmlands, cities
Australian Masked-Owl *Tyto novaehollandiae*	1	-	-	-	-	-	**Forest, savanna
Eurasian Eagle-Owl *Bubo bubo*	-	1	-	-	-	-	Forest, woodland, desert, farmlands
Little Owl *Athene noctua*	2	-	-	-	-	-	Steppes, stony semi-desert, farmlands, open woodland, towns
Rock Pigeon *Columba livia*	21	1	-	-	-	-	Cliffs, caves, ledges, eaves
Madagascar Turtle-Dove *Columba picturata*	4	1	-	1	-	-	Evergreen forest, second growth, scrub
European Turtle-Dove *Streptopelia turtur*	-	1	-	2	-	-	Open woodland, scrub, plains, gardens
Laughing Dove	2	2	-	-	-	-	Thorn scrub, oases, towns

Species							Habitat
Spotted Dove *Streptopelia senegalensis*	15	-	-	-	-	-	Woodland, farmlands, towns
Red Collared-Dove *Streptopelia chinensis*	-	2	-	1	-	-	Forest, woodland, farmlands
Red-eyed Dove *Streptopelia tranquebarica*	1	-	-	-	-	-	Broken forest, savanna, reedbeds, riparian forest, towns
Eurasian Collared-Dove *Streptopelia semitorquata*	5	-	-	7	-	-	Open woodland, scrub, brushy areas, desert, and around human habitation
Island Collared-Dove *Streptopelia decaocto*	-	5	-	2	-	-	Open country, fields, around human habitation
Emerald Dove *Streptopelia bitorquata*	1	-	-	2	-	-	**Forest, woodland
Common Bronzewing *Chalcophaps indica*	-	1	-	2	-	-	Dry open woodland, scrub, open country
Crested Pigeon *Phaps chalcoptera*	-	4	-	5	-	-	Arid plains, sparse dry woodland, ranches, generally near water
Zebra Dove *Geophaps lophotes*	13	4	-	4	-	-	Open forest, edge, towns, farmlands, savannas
Mourning Dove *Geopelia striata*	-	2	-	-	-	-	Open forest, farmlands, deserts, towns
White-winged Dove *Zenaida macroura*	2	3	-	-	-	-	Scrub, woodlands, farmlands, mangroves
Inca Dove *Zenaida asiatica*	-	1	-	-	-	-	Open country, scrub, arid brush, farmlands, towns
Common Ground-Dove *Columbina inca*	-	1	-	-	-	-	Farmlands, open country, arid brush
Caribbean Dove *Columbina passerina*	-	1	-	-	-	-	Forest undergrowth, open scrub, arid woodland
Weka *Leptotila jamaicensis*	7	1	-	1	NT	-	**Scrub, forest borders
Purple Swamphen *Gallirallus australis*	-	2	-	1	-	-	Marsh, tussock swamps, lake shores
Gough Moorhen *Porphyrio porphyrio*	1	-	-	-	VU	79	**Tussock grass, bushes
Common Moorhen *Gallinula comeri*	1	-	-	1	-	-	Marshes, swampy riparian bushes, reedbeds
Chatham Islands Snipe *Gallinula chloropus*	1	-	-	-	VU	209	**Forest floor, scrub
White-tailed Eagle *Coenocorypha pusilla*	-	2	-	-	NT	-	Rocky seacoasts, rivers, large lakes
Lammergeier *Haliaeetus albicilla*	-	1	-	-	-	-	Crags, ladges, gorges, around human habitation
Western Marsh-Harrier *Gypaetus barbatus*	1	-	-	-	-	-	Swamps, marsh, rice paddies, reed beds
Common Buzzard *Circus aeruginosus*	-	1	-	-	-	-	Woodland, forest, moors, heath, towns, farmlands
Chimango Caracara *Buteo buteo*	1	-	-	-	-	-	Open country, savanna, scrub, farmlands, often near water
Milvago chimango							

7. Are Unsuccessful Invaders Rare in Their Native Range? 141

Species							Habitat
Peregrine Falcon *Falco peregrinus*	-	2	-	-	-	-	Open country, tundra, moors, steppe, seacoasts, cliffs, cities
Cattle Egret *Bubulcus ibis*	3	-	-	1	-	-	Wet fields, marshes, swamps, pastures, grassland, often associated with large grazing mammals
Black-crowned Night-Heron *Nycticorax nycticorax*	1	-	-	-	-	-	Marshes, swamps, ponds, lakes, lagoons, mangroves
Greater Flamingo *Phoenicopterus ruber*	1	-	-	1	-	-	Salt lakes, brackish shallow lagoons, mudspits
Scarlet Ibis *Eudocimus ruber*	1	-	-	-	-	-	Coastal swamps, mangroves, lagoons, tidewater rivers
Turkey Vulture *Cathartes aura*	5	-	-	-	-	-	Forest, grasslands, desert, open country
Great Kiskadee *Pitangus sulphuratus*	1	-	-	-	-	-	Open woodland, savanna, towns, usually near water
Superb Lyrebird *Menura novaehollandiae*	1	-	-	-	-	-	**Forest, treefern gullies
Noisy Miner *Manorina melanocephala*	1	-	-	1	-	-	**Woodland, towns
Blue Magpie *Urocissa erythrorhyncha*	-	1	-	-	-	-	**Forest, woodland
Black-billed Magpie *Pica pica*	1	-	-	-	-	-	Open woodland, scrub, savanna
Eurasian Jackdaw *Corvus monedula*	-	2	-	1	-	-	Open woodland, farmland, towns
House Crow *Corvus splendens*	10	6	-	2	-	-	Towns, cities, near human habitation
New Caledonia Crow *Corvus moneduloides*	1	-	-	-	-	201	**Woodland, open areas
Rook *Corvus frugilegus*	1	-	-	-	-	-	Open country, pastures, farmlands, woodland
Greater Bird-of-Paradise *Paradisaea apoda*	1	-	-	-	-	179	*Forest
Australian Magpie *Gymnorhina tibicen*	3	1	-	2	-	-	Open woodland, savanna, farmlands
Black Drongo *Dicrurus macrocercus*	1	-	-	-	-	-	Woodland, open country, towns, marshes
Saddleback *Creadion carunculatus*	14	-	-	5	CD	-	*Forest
Blackbird *Turdus merula*	2	-	-	6	-	-	Woodland, undergrowth, thickets, towns, open areas
Island Thrush *Turdus poliocephalus*	1	-	-	-	-	-	Forest floor, edge, second growth
Song Thrush *Turdus philomelos*	2	-	-	4	-	-	Woodland, thickets, scrub, towns
Oriental Magpie-Robin *Copsychus saularis*	-	3	-	-	-	-	Farmlands, woodland, forest, but absent from desert
White-rumped Shama *Copsychus malabaricus*	2	-	-	-	-	-	Second growth thickets, bamboo, forest

Species							Habitat
Common Starling
Sturnus vulgaris	22	2	-	7	-	-	Farmlands, open country, woodland, towns
Common Myna
Acridotheres tristis	29	5	-	5	-	-	Open country, farmlands, cities
Jungle Myna
Acridotheres fuscus	3	1	-	2	-	-	Forest edge, clearings, farmlands
Crested Myna
Acridotheres cristatellus	3	1	-	-	-	-	**Open country, farmlands
Hill Myna
Gracula religiosa	2	-	-	4	-	-	Humid forest, woodland, second growth
Northern Mockingbird
Mimus polyglottus	4	-	-	6	-	-	Open scrub, forest edge, brush, farmland, towns
Tropical Mockingbird
Mimus gilvus	3	-	-	1	-	-	Open areas, thorn scrub, towns
Great Tit
Parus major	-	1	-	1	-	-	Open woodland, forest, towns, bamboo, mangroves
Varied Tit
Parus varius	-	1	-	-	-	-	**Open mixed woodland, forests
Red-whiskered Bulbul
Pycnonotus jocosus	7	2	-	-	-	-	Forest edge, second growth, towns
Red-vented Bulbul
Pycnonotus cafer	4	-	2	-	-	-	Scrub, second growth, woodland, towns
Sooty-headed Bulbul
Pycnonotus aurigaster	-	2	1	-	-	-	Scrub, second growth, woodland, towns
Japanese White-eye
Zosterops japonicus	2	-	-	-	-	-	Forest, woodland, farmlands, thickets
Christmas Island White-eye
Zosterops natalis	1	-	-	-	NT	188	Open country, edge of forest
Silvereye
Zosterops lateralis	3	-	-	4	-	-	Coastal shrubs, forest undergrowth, open scrub, towns, edge, second growth
Japanese Bush-Warbler
Cettia diaphone	1	-	-	-	-	-	Dense brush, undergrowth, tall grass
White-throated Laughingthrush
Garrulax albogularis	-	2	-	-	-	-	*Dense forest
White-crested Laughingthrush
Garrulax leucolophus	-	1	-	-	-	-	**Forest undergrowth, second
Greater Necklaced Laughingthrush
Garrulax pectoralis	-	2	-	-	-	-	**Forest undergrowth, second
Hwamei
Garrulax canorus	1	-	-	1	-	-	Thickets, bamboo, farmlands
White-browed Laughingthrush
Garrulax sannio	-	1	-	-	-	-	Open areas, farmlands, bamboo, scrub, thickets
Red-billed Leiothrix
Leiothrix lutea	2	2	-	4	-	-	Second growth, scrub, grass
Eurasian Skylark
Alauda arvensis	9	-	-	3	-	-	Grasslands, fields, tundra, marshy areas, sand dunes, clearings
House Sparrow
Passer domesticus	49	2	-	6	-	-	Commensal of man, towns, farmlands

Spanish Sparrow *Passer hispaniolensis*	3	-	-	-	-	-	Commensal of man, riparian woodland, bushes, fields, towns
Eurasian Tree Sparrow *Passer montanus*	8	1	-	3	-	-	Open woodland, plains, farms, towns, commensal of man
Hedge Accentor *Prunella modularis*	2	-	-	1	-	-	Shrubbery, thickets in woodland, heather, towns
Village Weaver *Ploceus cucullatus*	4	1	-	1	-	-	Savanna, forest clearings, swamps, towns
Black-headed Weaver *Ploceus melanocephalus*	-	1	-	-	-	-	Riparian vegetation, reedbeds, grass
Madagascar Red Fody *Foudia madagascarensis*	9	4	-	-	-	-	Brush, grass, forest
Yellow-crowned Bishop *Euplectes afra*	-	1	-	-	-	-	**Grassy marshes, riparian grassland
Red Bishop *Euplectes orix*	1	1	-	2	-	-	*Tall grass in moist areas
White-winged Widowbird *Euplectes albonotatus*	1	-	-	1	-	-	*Tall grass in acacia savanna
Blue-breasted Cordonbleu *Uraeginthus angolensis*	-	5	-	2	-	-	Savanna and woodland in bushes, thickets, towns
Red-cheeked Cordonbleu *Uraeginthus bengalus*	1	-	-	2	-	-	**Dry savanna, woodland in thickets
Blue-capped Cordonbleu *Uraeginthus cyanocephala*	-	1	-	-	-	-	*Dry thorn scrub
Black-tailed Waxbill *Estrilda perreini*	1	-	-	-	-	-	**Forest undergrowth, riparian thickets
Orange-cheeked Waxbill *Estrilda melpoda*	3	-	-	1	-	-	**Tall grass savanna, thickets
Black-rumped Waxbill *Estrilda troglodytes*	2	-	-	1	-	-	**Arid thorn scrub, farmlands
Common Waxbill *Estrilda astrild*	11	3	2	1	-	-	**Open grassland, edge, often near water
Red Avadavat *Amandava amandava*	3	5	1	7	-	-	Reedy swamps, grassland, scrub, farmlands
Green Avadavat *Amandava formosa*	-	1	-	-	VU	-	**Tall grass, scrub
Red-browed Firetail *Neochima temporalis*	3	1	-	3	-	-	Mangrove, forest, open areas, towns
Zebra Finch *Taeniopygia guttata*	1	-	-	4	-	-	Varied, including grass, shrubs, woodland (absent from dense forest)
Double-varred Finch *Taeniopygia bichenovii*	-	1	-	1	-	-	Varied; grass, Pandanus, scrub, farmlands
White-throated Silverbill *Lonchura malabarica*	-	2	-	-	-	-	Arid thorn savanna, tall grass, scrub, open woodland
Bronze Munia *Lonchura cucullata*	3	-	-	1	-	-	Savanna, scrub, farmlands
Magpie Munia *Lonchura fringilloides*	-	1	-	1	-	-	Riparian forest, clearings, bamboo
Javan Munia	-	1	-	-	-	-	**Second growth, farmlands

Species							Habitat
Scaly-breasted Munia *Lonchura leucogastroides*	6	-	-	3	-	-	Grass, scrub, farmlands
Black-headed Manakin *Lonchura punctulata*	2	2	1	1	-	-	Grass, scrub, reedbeds, farmlands
Chestnut-breasted Munia *Lonchura malacca*	3	-	-	3	-	-	Riparian grass, reedbeds, mangroves
Java Sparrow *Lonchura castensothorax* *Padda oryzivora*	22	4	5	10	VU	-	Scrub, mangroves, rice fields, towns
Pin-tailed Whydah *Vidua macroura*	1	-	-	2	-	-	*Savanna
Chaffinch *Fringilla coelebs*	3	-	-	3	-	-	Woodland, forest, farmlands
Island Canary *Serinus canaria*	2	-	-	7	-	120,s 69	Open woodland, scrub, sparse montane forest, farmlands
Cape Canary *Serinus canicollis*	1	-	-	3	-	-	Scrub, farmlands, grasslands, pine woodland
Yellow Canary *Serinus flaviventris*	2	-	-	-	-	-	**Arid scrub, coastal bush
White-rumped Seedeater *Serinus leucopygius*	-	1	-	-	-	-	*Savanna
Yellow-fronted Canary *Serinus mozambicus*	4	1	-	2	-	-	Savanna, woodlands, farmlands, in moister areas than S. atrogularis
European Greenfinch *Carduelis chloris*	6	-	-	2	-	-	**Open woodland, farmlands
Lesser Goldfinch *Carduelis psaltria*	-	2	-	-	-	-	Woodland edge, chaparral, riparian woodland, open country, farmlands
European Goldfinch *Carduelis carduelis*	7	2	-	5	-	-	Open country, woodland, farmlands, weedy areas
Common Redpoll *Carduelis flammea*	1	-	-	-	-	-	Open birch woodland, scrub, willows, tundra
House Finch *Carpodacus mexicanus*	2	-	-	-	-	-	Open woodland, farmlands, open country, arid scrub, pine-oak, savanna, towns
Nihoa Finch *Telespiza ultima*	1	-	-	1	VU	s138	**Rock-outcroppings, shrub-covered slopes
Laysan Finch *Telespiza cantans*	1	-	-	3	VU	216	Scaevola thickets, bunch-grass, low bushy areas
Yellowhammer *Emberiza citrinella*	1	-	-	3	-	-	Grasslands, bushes, farmlands
Cirl Bunting *Emberiza cirlus*	2	-	-	-	-	-	**Bushes in open country, woodland edge
Red-crested Cardinal *Paroaria coronata*	1	2	-	-	-	-	**Wet scrub, shrubbery
Yellow-billed Cardinal *Paroaria capitata*	1	-	-	-	-	-	Shrubbery in humid areas, forest edge, woodland
Crimson-backed Tanager	1	-	-	-	-	-	Thickets, scrub, humid forest

Species	A	B	C	D	Thr	EBA	Habitats
Ramphocelus dimidiatus							edge, open woodland, towns
Blue-grey Tanager *Thraupis episcopus*	1	1	-	1	-	-	Open woodland, forest edge, second growth, riverine woodland, towns, especially near water
Red-legged Honeycreeper *Cyanerpes cyaneus*	-	1	-	1	-	-	Humid forest edge, open woodland, second growth, towns
Common Duica-Finch *Diuca diuca*	1	-	-	-	-	-	Bushy hillsides, sand dunes, towns, arid gravelly hillsides
Saffron Finch *Sicalis flaveola*	3	-	-	-	-	-	Open grassland, savanna, open woodland, second growth, towns, farmlands
Grassland Yellow-finch *Sicalis luteola*	5	-	-	-	-	-	Grasslands, savanna, farmlands
Cuban Grassquit *Tiaris canora*	1	1	-	-	-	-	**Woodland, shrubbery bordering fields
Yellow-faced Grassquit *Tiaris olivacea*	1	-	-	1	-	-	Open grassy and shrubby areas, fields, second growth, forest edge, farmlands
Puerto Rican Bullfinch *Loxigilla portoricensis*	-	1	-	-	-	29, 30	Woodland, arid scrub, mangroves
Lesser Antilles Bullfinch *Loxigilla noctis*	-	1	-	-	-	29, 30	Shrubbery, forest undergrowth, towns
Northern Cardinal *Cardinalis cardinalis*	4	-	-	1	-	-	Thickets, fields, forest edge, riparian thickets, woodland, towns, arid scrub
Spot-breasted Oriole *Icterus pectoralis*	1	-	-	-	-	-	Open country with scattered trees, woodland edge, towns
Troupial *Icterus icterus*	4	-	-	6	-	-	*Woodland near rivers
Red-breasted Blackbird *Leistes militaris*	1	-	-	1	-	-	Bushy pastures, wet grasslands, swampy places
Western Meadowlark *Sturnella neglecta*	2	-	-	-	-	-	Grasslands, savanna, farmlands
Carib Grackle *Quiscalus lugubris*	2	-	-	1	-	-	Open woodland, farmlands, pastures, towns, arid scrub
Shiny Cowbird *Molothrus bonariensis*	15	-	-	-	-	-	Open woodland, farmlands, marshes, towns, second growth

Appendix II. Unsuccessfully introduced bird species. Details as in appendix I.

Species	A	B	C	D	Thr	EBA	Habitats
Greater Rhea *Rhea americana*	-	-	-	1	NT	-	**Grassy plains, open brush
Great Tinamou *Tinamus major*	-	-	-	1	-	-	*Humid, sometimes disturbed forest

Species							Habitat
Red-winged Tinamou *Rynchotus rufescens*	-	-	-	2	-	-	**Open country, grassland
Elegant Crested-Tinamou *Eudromia elegans*	-	-	-	1	-	-	Dry savanna, open woodland, dry steppes
Chestnut-winged Chacalaca *Ortalis garrula*	-	-	-	1	-	-	Tropical thickets, scrubby deciduous forest, dense second growth, arid scrub, riparian woodland, mangroves
Crested Guan *Penelope purpurascens*	-	-	-	1	-	-	Heavy humid forest, edge, riparian woodland
Black Curassow *Crax alector*	-	-	-	1	-	-	Heavy humid forest, edge, riparian thickets
Malleefowl *Leipoa ocellata*	-	-	-	6	VU	-	**Mallee and dry scrub of semi-arid zones
See-see Partridge *Ammoperdix griseogularis*	-	-	2	-	-	-	**Dry, rocky slopes with sparse vegetation, forest areas with cliffs
Sand Partridge *Ammoperdix heyi*	-	-	-	1	-	-	*Desert, especially in rocky areas
Himalayan Snowcock *Tetraogallus himalayensis*	-	-	2	-	-	-	**Steep stony slopes with sparse vegetation, alpine meadows
Arabian Partridge *Alectoris melanocephala*	-	-	1	-	-	-	*Rocky desert areas
Clapperton's Francolin *Francolinus clappertoni*	-	-	-	1	-	-	**Grasslands, savanna
Heuglin's Francolin *Francolinus icterorhynchus*	-	-	-	1	-	-	**Grasslands, savanna
Yellow-necked Francolin *Francolinus leucoscepus*	-	-	2	-	-	-	**Grasslands, farmlands
Stubble Quail *Coturnix pectoralis*	-	-	-	3	-	-	**Rank grassland, farmlands
Jungle Bush-Quail *Perdicula asiatica*	-	-	-	2	-	-	**Dry scrub and bush, second growth
Formosan Partridge *Arborophila crudigularis*	-	-	-	1	NT	149	*Forest
Crested Partridge *Rollulus roulroul*	-	-	-	1	-	-	**Forest, dense second growth
Temminck's Tragopan *Tragopan temmincki*	-	-	-	1	NT	-	Evergreen or mixed forest, dense rhododendron, bamboo
Himalayan Monal *Lophophorus impejanus*	-	-	-	2	-	-	Open forest, rhododendron, usually on rocky, broken, and precipitous slopes and in gorges
Green Jungle-fowl *Gallus varius*	-	-	1	-	-	-	*Forest
Grey Junglefowl *Gallus sonneratii*	-	-	-	1	NT	-	Forest, scrub, bamboo
Brown Eared-Pheasant *Crossoptilon mantchuricum*	-	-	-	1	VU	136	Bleak and rocky areas, shrubs, scrub, coarse grass, sparse and stunted
Copper Pheasant *Syrmaticus soemmerringii*	-	-	-	3	NT	-	Coniferous forest, especially Cryptomeria and cypress, dense

7. Are Unsuccessful Invaders Rare in Their Native Range? 147

Species							Habitat
							undergrowth of adjoining mixed forest, grassy hillsides
Rock Ptarmigan	-	-	-	1	-	-	*Tundra
Lagpous mutus							
Sage Grouse	-	-	-	2	-	-	*Sage grasslands and semi-desert
Centrocercus urophasianus							
Lesser Prairie-chicken	-	-	-	1	-	-	*Arid grasslands with shrubs and dwarf trees
Tympanuchus pallidicinctus							
Montezuma Quail	-	-	-	1	-	-	**Pine-oak forest, oak scrub
Cyrtonyx montezumae							
White-faced Whistling-Duck	-	-	1	-	-	-	Marshes, swamps, lagoons, rivers
Dendrocygna viduata							
Bean Goose	-	-	1	-	-	-	Tundra and taiga lakes, ponds, bogs, sluggish rivers, swamps, wet meadows
Anser fabilis							
Snow Goose	-	-	-	1	-	-	Open tundra, in migration marshes, wet prairies and flooded fields
Anser caerulescens							
Upland Goose	-	-	-	2	-	-	**Semi-arid grassy plains, open slopes
Chloephaga picta							
Orinoco Goose	-	-	-	1	NT	-	Rivers, marshes, lakes
Neochen jubata							
Spur-winged Goose	-	-	-	1	-	-	Lakes, rivers, marshes, swamps
Plectropterus gambensis							
Muscovy Duck	-	-	1	1	-	-	Forest streams, ponds, marshes, swamps
Cairina moschata							
Eurasian Wigeon	-	-	-	2	-	-	Marshes, lakes with emergent vegetation, open moors, in migration also in flooded fields and brackish coastal marshes
Anas penelope							
Blue-winged Teal	-	-	-	3	-	-	Marshes, ponds, sloughs, lakes, sluggish streams, in migration also streams and brackish marshes
Anas discors							
Common Pochard	-	-	1	1	-	-	Lakes, ponds, sluggish streams, especially with emergent vegetation, in migration also in brackish areas and estuaries
Aythya ferina							
Tufted Duck	-	-	-	2	-	-	Marshes, ponds, lakes, swamps, especially with emergent vegetation, in migration also rivers and brackish coastal areas
Aythya fuligula							
Painted Buttonquail	-	-	-	2	-	-	Forest undergrowth, scrub, rank grassland
Turnix varia							
Gang-gang Cockatoo	-	-	-	1	-	-	**Forest, woodland
Calocephalon fimbriatum							
Pink Cockatoo	-	-	-	1	NT	-	Grasslands, scrub, mulga, riverine forest, mallee
Cacatua leadbeateri							
Cockatiel	-	-	-	2	-	-	Open woodland, scrub, riverine forest, spinifex, farmlands
Nymphicus hollandicus							

148   *Chapter 7*

Blue-naped Parrot *Tanygnathus lucionensis*	-	-	-	1	EN	-	Forest, second growth, farmlands
Pale-headed Rosella *Platycercus adscitus*	-	-	-	1	-	-	Open woodland, scrub, riverine woodland, lantana thickets, farmlands, orchards
Kakapo *Strigops habroptilus*	-	-	-	4	EW	-	**Forest (esp. Nothofagus), scrubby snow tussock meadows
Vasa Parrot *Coracopsis vasa*	-	-	-	1	-	-	**Forest, savanna
Rosy-faced Lovebird *Agapornis roseicollis*	-	-	-	2	-	-	**Dry open country, savanna
Plum-headed Parakeet *Psittacula cyanocephala*	-	-	-	1	-	-	Open scrub, deciduous woodland, open forest, farmlands
Scarlet Macaw *Ara macao*	-	-	-	2	-	-	Forest edge, open woodland, savanna, farmlands
Red-lored Amazon *Amazona autumnalis*	-	-	-	1	-	-	Forest, mangroves, second growth, woodland, pine savanna, farmlands
Orange-winged Amazon *Amazona amazonica*	-	-	-	1	-	-	Forest edge, woodland, savanna, mangroves, farmlands
Edible-nest Swiftlet *Collocalia fuciphaga*	-	-	-	2	-	-	Open country, coasts and towns, breeding in caves
Tawny Owl *Strix aluco*	-	-	-	1	-	-	Forest, woodland, towns
Morepork *Ninox novaeseelandiae*	-	-	-	2	-	-	**Forest, farmlands
Common Wood-Pigeon *Columba palumbus*	-	-	-	1	-	-	Woodland, suburbs, towns
Namaqua Dove *Oena capensis*	-	-	-	2	-	-	Dry open country, farms, villages
Spinifex Pigeon *Geophaps plumifera*	-	-	-	2	-	-	*Dry rocky hills, usually associated with spinifex grass
Partridge Pigeon *Geophaps smithii*	-	-	-	2	NT	187	*Sparsely-wooded savanna
Diamond Dove *Geopelia cuneata*	-	-	-	5	-	-	Savanna waterholes, open woodland, mulga scrub, gum creeks
Bar-shouldered Dove *Geopelia humeralis*	-	-	-	2	-	-	Gallery forest, riparian scrub, mangroves, around human habitation
Wonga Pigeon *Leucosarcia melanoleuca*	-	-	-	3	-	-	*Humid forest
White-tipped Dove *Leptotila verreauxi*	-	-	-	1	-	-	Dry open woodland, forest edge, second growth, farmlands
Ruddy Quail-Dove *Geotrygon montana*	-	-	-	1	-	-	Humid forest undergrowth, second growth, woodland
Blue-headed Quail-Dove *Starnoenas cyanocephala*	-	-	-	3	EN	-	*Forest undergrowth
Nicobar Pigeon *Caloenas nicobarica*	-	-	-	2	NT	-	Bushes, thick forest, mangroves

7. Are Unsuccessful Invaders Rare in Their Native Range?

Species							Habitat
Luzon Bleeding-heart *Gallicolumba luzonica*	-	-	-	3	NT	151	*Forest
Pink-headed Imperial-Pigeon *Ducula rosacea*	-	-	-	1	-	162,	**Forest, woodland
Brolga *Grus rubicundus*	-	-	-	1	-	-	Swamps, marshes, flooded fields
Demoiselle Crane *Grus virgo*	-	-	-	1	-	-	Plains, steppes, desert, fields
Corn Crake *Crex crex*	-	-	-	1	VU	-	**Grassland, riparian thickets
Laysan Crake *Porzana palmeri*	-	-	-	4	EX	216	**Grass, reedbeds
Pallas's Sandgrouse *Syrrhaptes paradoxus*	-	-	-	2	-	-	*Arid sandy steppes (associated with *Artemisia absinthium*)
Pin-tailed Sandgrouse *Pterocles alchata*	-	-	-	1	-	-	Arid plains, bleak plateaus, dry mud flats
Chestnut-bellied Sandgrouse *Pterocles exustus*	-	-	3	3	-	-	*Sparse bushy arid land
Black-bellied Sandgrouse *Pterocles orientalis*	-	-	-	1	-	-	Arid stony stretches, fields
Eurasian Golden-Plover *Pluvialis apricaria*	-	-	-	3	-	-	Moors, bogs, swampy heath, wet and mossy tundra; in migration pastures, grasslands, fields, mudflats and estuaries
Grey Plover *Pluvialis squatarola*	-	-	-	2	-	-	Tundra; in migration mudflats, beaches, wet savanna, ponds, lakes, flooded fields and pastures
Northern Lapwing *Vanellus vanellus*	-	-	-	5	-	-	Open fields, pastures, wet meadows, bogs, lakes, in migration farmlands, seacoasts and mudflats
Western Gull *Larus occidentalis*	-	-	-	1	-	-	Rocky seacoasts, coastal cliffs, in winter also bays and estuaries
Silver Gull *Larus novaehollandiae*	-	-	-	1	-	-	Seacoasts, islands, large lakes
Great Cormorant *Phalacrocorax carbo*	-	-	-	1	-	-	Lakes, rivers, seacoasts, marshes
Guanay Cormorant *Phalacrocorax bougainvillei*	-	-	1	-	-	-	**Seacoasts, islands
Nankeen Night-heron *Nycticorax caledonicus*	-	-	-	1	-	-	Swamps, rivers, marshes, lakes, flooded areas, mangroves
Brown Pelican *Pelecanus occidentalis*	-	-	1	-	-	-	Seacoasts, estuaries, bays, islands
King Penguin *Aptenodytes patagonicus*	-	-	-	1	-	-	**Pelagic, breeding in tussock grass thickets
Bush Wren *Xenicus longipes*	-	-	-	1	EX	-	*Forest
Superb Fairywren *Malurus cyaneus*	-	-	-	1	-	-	Bushes, woodland, towns
American Crow	-	-	-	1	-	-	Open country, woodland,

Corvus brachyrhynchos							farmlands, orchards, towns
Large-billed Crow	-	-	-	1	-	-	Forest, edge, farmlands, towns
Corvus macrorhynchos							
Pied Crow	-	-	-	1	-	-	*Open country; often near human
Corvus albus							habitation; absent from extreme desert and dense forest
Grey Currawong	-	-	-	1	-	-	Forest, woodland, mallee
Strepera versicolor							
Willie-wagtail	-	-	-	1	-	-	Open areas with perches, often
Rhipidura leucophrys							near water, forest edge, farmlands
Magpie Lark	-	-	-	3	-	-	Open woodland, wet areas,
Grallina cyanoleuca							pastures, towns
White-throated Dipper	-	-	-	1	-	-	*Rapidly flowing rocky streams
Cinclus cinclus							
Western Bluebird	-	-	-	1	-	-	Open woodland, forest edge,
Sialia mexicana							farmlands, savanna
Hermit Thrush	-	-	-	1	-	-	Mixed forest or pure conifers,
Catharus guttatus							wooded bogs, dry sandy pine woodland, second growth
Red-legged Thrush	-	-	-	1	-	-	Forest, woodlands, towns,
Turdus plumbeus							thickets
American Robin	-	-	-	2	-	-	Forest, edge, open woodland,
Turdus migratorius							pastures, towns, savanna, farmlands
Narcissus Flycatcher	-	-	-	1	-	-	Forest, woodland, generally near
Ficedula narcissina							water
Blue-and-white Flycatcher	-	-	-	1	-	-	Forest, woodland, cliffs near
Cyanoptila cyanomelana							streams
European Robin	-	-	-	4	-	-	Shrubbery, woodland, forest,
Erithacus rubecula							towns
Japanese Robin	-	-	-	1	-	-	*Dense undergrowth, often
Erithacus akahige							riparian
Ryukyu Robin	-	-	-	1	NT	148	*Dense undergrowth, often
Erithacus komadori							riparian
Common Nightingale	-	-	-	4	-	-	**Undergrowth, thickets
Luscinia megarhynchos							
Seychelles Magpie-Robin	-	-	-	1	CR	100	**Woodland, towns
Copsychus sechellarum							
White-headed Starling	-	-	-	1	NT	125, 126	Grasslands, farmlands, second growth
Sturnus erythropygius							
Rosy Starling	-	-	-	1	-	-	**Open plains, rocky areas
Sturnus roseus							
Black-collared Starling	-	-	-	1	-	-	Open country, scrub, farmlands
Sturnus nigricollis							
Blue Tit	-	-	-	2	-	-	Woodland, bushes, towns
Parus caeruleus							
Mauritius Bulbul	-	-	1	-	VU	102	Forest, second growth, woodland
Hypsipetes olivaceus							
Black-throated Laughingthrush	-	-	-	1	-	-	Forest undergrowth, dense scrub,
Garrulax chinensis							thickets, bamboo, grass

7. Are Unsuccessful Invaders Rare in Their Native Range? 151

Species							Habitat
Blackcap *Sylvia atricapilla*	-	-	-	2	-	-	Forest with undergrowth, edge, open woodland, town, orchards, wooded steppes
Greater Whitethroat *Sylvia communis*	-	-	-	2	-	-	Low shrubby growth, scrub, steppe, open woodland, edge of farmlands, swamp edge, shrubby areas in grasslands
Mongolian Lark *Melanocorypha mongolica*	-	-	-	3	-	-	*High plains
Wood Lark *Lullula arborea*	-	-	-	2	-	-	**Grassland with small trees, woodland edges
Scaly Weaver *Sporopipes squamifrons*	-	-	-	1	-	-	*Dry thorn scrub
Cape Weaver *Ploceus capensis*	-	-	-	1	-	-	Trees, bushes, reedbeds, usually near water
Southern Masked-Weaver *Ploceus velatus*	-	-	-	1	-	-	**Acacia woodland, dry savanna
Baja Weaver *Ploceus philippinus*	-	-	-	2	-	-	Grassland, second growth, farmlands, reedbeds
Asian Golden Weaver *Ploceus hypoxanthus*	-	-	-	1	NT	-	Riparian scrub, reedbeds, farmlands
Seychelles Fody *Foudia sechellarum*	-	-	-	1	-	100	*Forest
Long-tailed Widowbird *Euplectus progne*	-	-	-	1	-	-	**Open grassland, dry short grass
Green-winged Pytilla *Pytilia melba*	-	-	-	1	-	-	**Bushes, thorn scrub in savanna
Red-billed Firefinch *Lagonosticta senegala*	-	-	1	1	-	-	**Savanna grassland, farmlands
Common Grenadier *Uraeginthus granatina*	-	-	-	1	-	-	*Arid thorn scrub
Swee Waxbill *Estrilda melanotis*	-	-	-	1	-	-	Grassy clearings, thickets, scrub
Black-cheeked Waxbill *Estrilda erythronotos*	-	-	-	1	-	-	*Dry thornbrush thickets
Zebra Waxbill *Amandava subflava*	-	-	-	2	-	-	*Savanna grassland
Diamond Firetail *Stagonopleura guttata*	-	-	-	7	-	-	Savanna woodland, mallee, usually near water
Star Finch *Neochima ruficauda*	-	-	-	1	VU	-	**Grassy riparian vegetation, swamps
Plum-headed Finch *Neochima modesta*	-	-	-	1	-	-	**Savanna woodland undergrowth, riparian thickets
Long-tailed Finch *Poephila acuticauda*	-	-	-	1	-	-	Savanna woodland, Pandanus, near wayer
Pin-tailed Parrotfinch *Erythrura prasina*	-	-	-	1	-	-	Forest, scrub, bamboo
Blue-faced Parrotfinch *Erythrura trichroa*	-	-	-	1	-	-	**Forest edge, mangroves
Red-throated Parrotfinch	-	-	-	1	-	201	Second growth, farmlands, open

Erythrura psittacea							woodland, edge
Red-headed Parrotfinch	-	-	-	1	-	203	**Open forest, second growth
Erythrura cyaneovirens							
Gouldian Finch	-	-	-	1	EN	-	Savanna, reedbeds, mangroves,
Chloebia gouldiae							usually near water
White-rumped Manakin	-	-	-	1	-	-	Second growth, farmlands, edge
Lonchura striata							
Village Indigobird	-	-	-	1	-	-	**Savanna, brushy areas
Vidua chalybeata							
Queen Whydah	-	-	-	1	-	-	*Thorn savanna
Vidua regia							
Eastern Paradise-Whydah	-	-	-	1	-	-	*Open woodland
Vidua paradisaea							
Brambling	-	-	-	2	-	-	**Open birch-conifer woodland,
Fringilla montifringilla							riparian willows
Southern Yellow-rumped	-	-	-	1	-	-	Dry open areas, scattered trees,
Seedeater							savanna
Serinus atrogularis							
Eurasian Siskin	-	-	-	5	-	-	**Coniferous woodland, birch
Carduelis spinus							and alder thickets
American Goldfinch	-	-	-	2	-	-	Riparian woodland, fields, open
Carduelis tristis							woodland, forest edge, farmlands
Twite	-	-	-	1	-	-	*Open arid, stony ground with
Carduelis flavirostris							sparse bushes, grass
Eurasian Linnet	-	-	-	4	-	-	**Open country with sparse
Carduelis cannabina							vegetation, farmlands
Parrot Crossbill	-	-	-	1	-	-	*Coniferous forest, especially
Loxia pytyopsittacus							pines
Eurasian Bullfinch	-	-	-	6	-	-	Coniferous and mixed forest,
Pyrrhula pyrrhula							woodland, farmlands
Hawfinch	-	-	-	1	-	-	Deciduous and mixed woodland,
Coccothraustes coccothraustes							farmlands, scrub
Ortolan Bunting	-	-	-	2	-	-	Open country with scattered
Emberiza hortulana							bushes, scrub, towns, farmlands,
							woodland edge
Reed Bunting	-	-	-	1	-	-	Reedbeds, rushes, riparian
Emberiza schoeniclus							thickets
Rufous-collared Sparrow	-	-	-	1	-	-	Open situations with scattered
Zonotrichia capensis							bushes, shrubby hillsides,
							thickets, farmlands, humid forest
							edge, open woodland, towns
Yellow Cardinal	-	-	-	1	EN	-	*Shrubbery
Gubernatrix cristata							
Red-cowled Cardinal	-	-	-	2	-	-	Open forest, edge, second growth
Paroaria dominicana							
Red-capped Cardinal	-	-	-	1	-	-	Open forest, edge, second
Paroaria gularis							growth, scrub, especially near
							water
White-lined Tanager	-	-	-	1	-	-	Humid forest edge, second
Tachyphonus rufus							growth, thickets, deciduous

							forest, scrub, open woodland, savanna
Summer Tanager *Piranga rubra*	-	-	-	1	-	-	Deciduous forest, open woodland, pine-oak, riparian woodland, pine woodland, towns
Scarlet Tanager *Piranga olivacea*	-	-	-	1	-	-	Deciduous forest, woodland, parks
Silver-beaked Tanager *Ramphocelus carbo*	-	-	-	1	-	-	Bushes, second growth, forest edge, open woodland, towns, especially near water
Brazilian Tanager *Ramphocelus bresilius*	-	-	-	1	-	-	Forest edge, second growth scrub, bushes, swampy woodland, gardens, especially near water
Scarlet-rumped Tanager	-	-	-	1	-	-	Thickets, second growth, scrub, humid forest edge, towns, especially near
Antillean Euphonia *Euphonia musica*	-	-	-	1	-	-	Woodland, forest, edge, scrub
Golden Tanager *Tangara arthus*	-	-	-	1	-	-	**Humid forest, edge
Masked Tanager *Tangara nigrocincta*	-	-	-	1	-	-	Humid forest edge, second growth, open woodland
White-collared Seedeater *Sporophila torqueola*	-	-	-	1	-	-	Bushy and weedy areas, open woodland and scrub, farmlands, savanna
Indigo Bunting *Passerina cyanea*	-	-	-	1	-	-	Deciduous forest edge, open woodland, second growth, scrub, farmlands
Painted Bunting *Passerina ciris*	-	-	-	1	-	-	Brushy areas, scrub, riparian thickets, forest edge, second growth
Orange-breasted Bunting *Passerina leclancherii*	-	-	-	2	-	-	Deciduous forest, arid scrub, brush, old fields
Morioche Oriole *Icterus chrysocephalus*	-	-	-	1	-	-	**Forest, open woodland
Yellow-hooded Blackbird *Agelaius icterocephalus*	-	-	-	1	-	-	**Marshes, grassy swamps
Pampas Meadowlark *Sturnella militaris*	-	-	-	1	-	-	*Grasslands

REFERENCES

Bazzaz, F. A. 1986. Life history of colonizing plants: some demographic, genetic, and physiological features. In: Ecology of Biological Invasions of North America and Hawaii (H. A. Mooney, and J. A. Drake eds.), Springer-Verlag, New York, Pp: 96–110.

Beck, B. B., L.G. Rapaport, Stanely, M.R. Price, and A.C. Wilson. 1994. Reintroduction of captive-born animals. In: Creative Conservation (P. J. S. Olney, G. M. Mace, and A. T. C. Feistner eds.), Champan & Hall, London, Pp: 264–286.

Collar, N. J., M.J. Crosby, and A.J. Stattersfield. 1994. Birds to Watch 2. The World List of Threatened Birds. BirdLife Conservation Series No. 4. BirdLife International, Cambridge.

Diamond, J., S.L. Pimm, M.E. Gilpin, and M. LeCroy. 1989. Rapid evolution of character displacement in Myzomelid Honeyeaters. American Naturalist 134:675–708.

Duncan, R. P. 1997. The role of competition and introduction effort in the success of Passeriform birds introduced into New Zealand. American Naturalist 149:903–915.

Ehrlich, P. R. 1986. Which animal will invade? In: Ecology of Biological Invasions of North America and Hawaii (H. A. Mooney, and J. A. Drake eds.), Springer-Verlag, New York, Pp: 79–92.

Ehrlich, P. R. 1989. Attributes of invaders and the invading process: vertebrates. In: Biological Invasions: a Global Perspective (J. A. Drake, H. A. Mooney, F. di Castri, R. H. Groves, F. J. Kruger, M. Rejmánek, and M. Williamson eds.), SCOPE 37. John Wiley & Sons, Chichester, Pp: 315–328.

Franklin, J., and D.W. Steadman. 1991. The potential for conservation of Polynesian birds through habitat mapping and species translocation. Conservation Biology 5:506–521.

Garrett, K. L. 1993. Canary-winged Parakeets: the southern California perspective. Birding 25:430–431.

Gaston, K. J. 1994. Rarity. Chapman & Hall, London.

Gilpin, M. 1990. Ecological prediction. Science 248:88–89.

Goerck, J. M. 1997. Patterns of rarity in the birds of the Atlantic forest of Brazil. Conservation Biology 11:112–118.

Goodwin, B. J., A.J. McAllister, and L. Fahrig. 1999. Predicting invasiveness of plant species based on biological information. Conservation Biology 13:422–426.

Green, R. E. 1997. The influence of numbers released on the outcome of attempts to introduce exotic bird species to New Zealand. Journal of Animal Ecology 66:25–35.

Griffith, B., J.M. Scott, J.W. Carpenter, C. and Reed. 1989. Translocation as a species conservation tool: status and strategy. Science 245:477–480.

Holt, R. D., J.H. Lawton, K.J. Gaston, and T.M. Blackburn. 1998. On the relationship between range size and local abundance: back to basics. Oikos 78:183–190.

Lever, C. 1987. Naturalized Birds of the World. Longman Scientific and Technical, New York.

Lockwood, J. L. 1999. Using taxonomy to predict success among introduced avifauna: relative importance of transport and establishment. Conservation Biology 13:560–567.

Lockwood, J. L., T.M. Brooks, and M.L. McKinney. In press. Taxonomic homogenization of the global avifauna. Animal Conservation.

Lodge, D. M. 1993. Biological invasions: lessons for ecology. Trends in Ecology and Evolution 8:133–137.

Long, J. 1981. Introduced Birds of the World. David & Charles, London.

Lövei, G. L. 1997. Global change through invasion. Nature 388:627.

Kattan, G. H. 1992. Rarity and vulnerability: the birds of the Cordillera Central of Colombia. Conservation Biology 6:64–70.

Moulton, M. P., and S.L. Pimm, S. L. 1986. Species introductions to Hawaii. In: Ecology of Biological Invasions of North America and Hawaii (H. A. Mooney, and J. A. Drake eds.), Springer-Verlag, New York, Pp: 231–249

Moulton, M. P., and J.G. Sanderson. 1997. Predicting the fates of passeriform introductions on oceanic islands. Conservation Biology 11:552–558.

Monroe, B. L., Jr., and C.G. Sibley. 1993. A World Checklist of Birds. Yale University Press, Newhaven.

Olson, S. L., and H.F. James. 1986. The holotype of the Laysan Finch Telespiza cantans Wilson (Drepanidini). Bulletin of the British Ornithologists' Club 106:84–86.

Pimm, S. L. 1989. Theories of predicting success and impact of introduced species. In: Biological Invasions: a Global Perspective (J. A. Drake, H. A. Mooney, F. di Castri, R. H. Groves, F. J. Kruger, M. Rejmánek, and M. Williamson eds.), SCOPE 37, John Wiley & Sons, Chichester, Pp: 351–367

Pitelka, F. A. 1981. The condor case: an uphill struggle in a downhill crush. Auk 98:634–635.

Rabinowitz, D. 1981. Seven forms of rarity. In: The Biological Aspects of Rare Plant Conservation (H. Synge ed.), John Wiley & Sons, Chichester, Pp: 205–217.

Rabinowitz, D., S. Cairns, and T. Dillon. 1986. Seven forms of rarity and their frequency in the flora of the British Isles. In: Conservation Biology: the Science of Scarcity and Diversity (M. E. Soulé ed.), Sinauer Associates, Sunderland, Pp: 182–204.

Roy, J., M.L. Navas, and L. Sonié. 1991. Invasion by annual brome grasses: a case study challenging the homocline approach to invasions. In: Biogeography of Mediterranean Invasions (R. H. Groves, and F. di Castri eds.), Cambridge University Press, Cambridge, Pp: 207–224.

Russell, G. J., T.M. Brooks, M.L. McKinney, and C.G. Anderson. 1998. Present and future taxonomic selectivity in bird and mammal extinctions. Conservation Biology 12:1365–1376.

Scott, J. K., and F.D. Panetta. 1993. Predicting the Australian weed status of southern African plants. Journal of Biogeography 20:87–93.

Sibley, C. G., and B.L. Monroe, Jr. 1990. Distribution and Taxonomy of Birds of the World. Yale University Press, Newhaven.

Sibley, C. G., and B.L. Monroe, Jr. 1993. A Supplement to Distribution and Taxonomy of Birds of the World. Yale University Press, Newhaven.

Simberloff, D., and W. Boecklen. 1991. Patterns of extinction in the introduced Hawaiian avifauna: a reexamination of the role of competition. American Naturalist 138:300–327.

Smith, P. W., and S.A. Smith. 1993. An exotic dilemma for birders: the Canary-winged Parakeet. Birding 25:426–430.

Spiegel, M. R. 1994. Theory and Problems of Statistics. Second edition. McGraw-Hill, Inc., New York.

Stattersfield, A. J., M.J. Crosby, A.J. Long, and D.C. Wege. 1998. Endemic Bird Areas of the World. Priorities for Biodiversity Conservation. BirdLife Conservation Series No. 7. BirdLife International, Cambridge.

Toone, W. D., and M.P. Wallace. 1994. The extinction in the wild and reintroduction of the California condor (*Gymnogyps californicus*). In: Creative Conservation (P. J. S. Olney, G. M. Mace, and A. T. C. Feistner eds.), Champan & Hall, London, Pp: 411–419.

Veltman, C. J., S. Nee, and M.J. Crawley. (1996) Correlates of introduction success in exotic New Zealand birds. American Naturalist 147:542–557.

Vitousek, P. M., C.M. D'Antonio, L.L. Loope, and R. Westbrooks. 1996. Biological invasions as global change. American Scientist 84:468–478.

Williamson, M. 1996. Biological Invasions. Chapman & Hall, London.

Chapter 8

A Geographical Perspective on the Biotic Homogenization Process: Implications from the Macroecology of North American Birds

Brian A. Maurer[1], Eric T. Linder[2], and David Gammon[2]
[1]*Department of Fisheries and Wildlife & Department of Geography, Michigan State University, East Lansing, MI 48821, USA*
[2]*Department of Zoology, Brigham Young University, Provo, UT 84604, USA*

1. INTRODUCTION

The major concern regarding the current trend towards biotic simplicity is that it is occurring with unprecedented rapidity across the entire globe. Geographic ranges of many species are collapsing (Lomolino and Channel 1995, 1998; Channel and Lomolino 1999a), while species that are favored by ecological conditions associated with modern human-dominated ecosystems are expanding rapidly (see Bright 1996 and references therein). Invading species often establish viable populations in extremely disturbed conditions and persist by continual colonization of newly disturbed area (Bright 1996, Williams and Meffe 1999), and in some cases by altering ecosystem processes such as fire frequency to favor their persistence (e.g., Cronk and Fuller 1995). Often vigorous colonizers in early succession in their native ecosystems, invasive species put most of their energy into reproduction, are generally less well adapted to specific environments, and tend to be able to tolerate a wide variety of conditions (Williams and Meffe 1999).

In this chapter, we examine the geographic nature of the processes that lead to simplification and homogenization of natural communities. We focus on two major processes: the shrinkage and expansion of geographic ranges. A major product of human alteration of natural ecosystems is their conversion to landscapes that cannot support populations of native species previously occupying them (Turner et al. 1990). As the number of altered landscapes within a species' geographic range increases, the size of the geographic range must begin to contract. Exactly how this contraction occurs, however, is not clear, although the pattern that results from sustained shrinkage suggests that species tend to persist in regions on the periphery of their historical range (Lomolino and Channell 1995, 1998; Channel and Lomolino 1999b). Weedy species that have adaptations that allow them to thrive in human dominated landscapes expand their geographic ranges as human activities spread across a continent. Early models that sought to describe this process of expansion are probably too simplistic to account for the rate and direction of spread. We clearly need to know more about how the ecological adaptations of a weedy species translate into patterns of range expansion and stabilization.

Our views on the contraction and expansion of geographic ranges are intended to explicitly connect the scale of habitat use and population dynamics studied by ecologists with patterns of geographic range dynamics documented by biogeographers. To be useful, such views much rigorously interface with both disciplines, and hence, must be explicitly mutliscalar. We show how the paradigm of geographic range structure accomplishes this, and how macroecological data can be used to test predictions made under that paradigm.

2. THE GEOGRAPHIC RANGE STRUCTURE PARADIGM

Since homogenization ultimately occurs at a geographic scale, it is important to understand how geographic ranges in general are structured. Species are not uniformly distributed within the boundaries of their geographic ranges, but occur with high abundances in more centrally located portions of their ranges, and at lower abundances near the periphery (Hengeveld and Haeck 1981, 1982; Brown 1984; Maurer and Villard 1994; Brown et al. 1995). Peripheral populations have lower abundances because ecological conditions for them are more severe in those locations (Brown 1984). Consequently, peripheral populations are demographically less stable (Maurer and Brown 1989, Curnutt et al. 1996, Maurer 1999). In a geographic range there may be several regions of high abundance, and few ranges are symmetrical, with abundance declining away from central regions approximately equally in all directions. The underlying demography that causes these patterns is a consequence of the suite of ecological adaptations that form the niche of a species.

Different species have different niches because evolution has proceeded differently in each species. These differences among species are played out in geographic space such that each species' range has a different size and shape. These different sizes and shapes indicate how species respond to different patterns of resource availability across a continent (Brown and Maurer 1989; Maurer 1994, 1999; Brown 1995, Linder et al. in press). Species with large geographic ranges tend to be abundant, and have ranges that are less fragmented than species with small ranges (Brown and Maurer 1987, Maurer and Nott 1998, Maurer 1999). Species with small ranges, all else being equal, should be more susceptible to extinction because their populations cover less of their range.

Given this static picture of geographic ranges, what is known about how geographic ranges change over time? There is surprisingly little information regarding geographic range dynamics. The best information is from invasions of exotic species on continents (Hengeveld 1989, Williamson 1996). The spread of an invading species has received considerable theoretical attention (Skellam 1951, van den Bosch et al. 1992, Veit and Lewis 1996). These theoretical treatments have assumed more or less that the geographic space into which a species is spreading is essentially uniform in ecological conditions. This assumption, however, must be rejected in most cases, because what we know about geographic range structure indicates that ecological conditions in geographic space are exceedingly non-uniform with respect to the niche requirements of a species. Hence, although estimates of parameters for these range expansion models often seem to give reasonable fits to range dynamical data, there is no convincing scientific evidence that they describe the population mechanisms that cause ranges to expand.

In what follows, we examine data from summer and winter bird surveys of North America to better understand range expansions and contractions. First, we examine differences in geographic range structure between species with populations that are increasing and decreasing over short time scales (10 - 30 years) in order to examine how geographic ranges fluctuate over time. Our analysis does not examine the specific reasons why populations of each species are changing, rather, we seek to identify common "signatures" of range structure changes in response to environmental changes that indicate common underlying population dynamical processes. Second, we examine in detail the expansion of two species of birds introduced into eastern North America. European Starlings (*Sturnus vulgaris*) were introduced into the New York City area in the 1890's, and have expanded to fill the continent since then (Wing 1941). The invasion of Starlings in North America has been analyzed previously under the assumption of a uniform continent, we will demonstrate that the expansion dynamics of Starlings was in fact exceedingly biased in certain directions, implying extensive ecological heterogeneity of the continent with respect to the Starling's niche. A more recent example of range expansion is that of the House Finch

(*Carpodacus mexicanus*), released in the same vicinity as Starlings, but more recently in the early 1960's. The dynamics of this event have been documented by recent continent wide censuses. Here again we will show that the spread of the species is non-uniform in geographic space. We also show how the dynamics of range spread have occurred indicating that there is a leading edge of rapid population growth along the expanding border of the range, followed by relative stability in regions of high abundance. After presenting these analyses, we discuss the implications of the patterns for the processes of range expansion and contraction that accompany biological homogenization.

3. DIFFERENCES IN RANGE STRUCTURE BETWEEN INCREASING AND DECREASING SPECIES

In order to develop a picture of how geographic ranges change over time, we compiled data from the North American Breeding Bird Survey (BBS). The BBS has been extensively analyzed and described elsewhere (Robbins et al. 1986, Maurer 1994). Briefly, the BBS is a large census effort conducted annually at 4000 routes distributed across North America. Annual BBS censuses have been conducted since 1966. From each census, the number of birds counted along a 40km transect is obtained, and this is the basic datum that we use to examine geographic range dynamics. These data were used to construct maps estimating patterns of spatial variation in local population abundance (Maurer 1994). A variety of measures of various aspects of the structure of geographic ranges can be calculated with such data. We focus on two informative measures for which we were able to obtain estimates over three five-year time intervals: 1966-1970, 1971-1975, and 1976-1980.

The first measure is the box dimension of the geographic range (Maurer 1994). The box dimension is a number between one and two that reflects the degree of jaggedness and fragmentation of a geographic range. When the box dimension is close to two, the range is nearly two-dimensional, has few internal gaps, and has a relatively smooth border. When the box dimension is near one, the geographic range is highly fragmented, has many internal gaps, and has a very jagged border. A smaller box dimension implies that a geographic range has a high ratio of perimeter to area. This suggests that a large fraction of populations are close to the range boundary and experience poor conditions (Maurer and Nott 1998).

Our second measure of geographic range fragmentation is the exponent of the power relationship between the standard deviation of population

8. A Geographic Perspective on Biotic Homogenization

density and average abundance. Taylor (Taylor 1961, Taylor and Taylor 1977) first studied this exponent. Let s and x be the standard deviation and average, respectively, of abundance over time on a single BBS route, then Taylor showed that

$$s = ax^b$$

Taylor's exponent, b, has a unique interpretation with respect to geographic ranges (Maurer 1994, Curnutt et al. 1996). When $b < 1$, then populations at the periphery of the range have higher coefficients of variation, hence are relatively more variable, than populations near the center of the range. Conversely, when $b > 1$, populations at the periphery are relatively less variable than central populations. Estimates of b can be obtained by plotting standard deviations and means from many sites (e.g., BBS routes) on logarithmic axes and calculating the slope of the resulting bivariate relationship.

For each measure, we selected species for which we had appropriate data to calculate the measure. We selected only species that showed an overall decline or increase across their entire geographic range from 1966 to 1980 as given by the US Geological Survey, Biological Resources Division (Sauer et al. 1996). Some species show changes in abundance in some parts of their ranges but not in others (James et al. 1996, Villard and Maurer 1996). We left such species out of our analysis because we required only species that showed clear cut increases or decreases across their entire ranges.

Each measure was analyzed using repeated measures analysis of variance (RMANOVA). The between-subjects factor was the population trend of a species (increasing or decreasing) and the within-subjects factor was time interval (1966-1970, 1971-1975, 1976-1980). Box dimensions of species that showed declining population trends were higher (implying less fragmentation) than those with increasing trends, but this was due to the fact that in our sample of species with range declines, there were several species with large ranges. There is a negative correlation between range size and the degree of fragmentation of the range boundary (Maurer and Nott 1998). To remove the range size effect from the analysis, we conducted the RMANOVA of box dimensions with geographic range size as a covariate. For both box dimensions and Taylor's exponents, there was a significant interaction between population trend and time interval (Table 1).

Species with decreasing population trends showed a decline in average box dimension, implying that such species, on the average, had geographic ranges that became more fragmented over time (Fig 1). The significant interaction was reflected by the tendency of increasing species to show the opposite trend: their ranges became less fragmented with time (Fig. 1).

Table 1. F-statistics for RMANOVA tests of differences between species of North American passerine birds that had increasing and decreasing population trends throughout their ranges from 1966 to 1980. The trend effect refers to whether a species had an increasing or decreasing population trend. The time effect refers to the following five-year periods: 1966-1970, 1971-1975, 1975-1980. A significant interaction effect (trend x time) indicates that species with increasing populations showed a different temporal pattern than those that had decreasing populations. Analysis for the box dimension variable was adjusted for differences in geographic range sizes.

Effect	Box dimension (df)	Taylor parameter (df)
Trend	10.32* (1, 33)	1.33 (1, 53)
Time	2.51 (2, 32)	7.73* (2, 52)
Trend x time	5.52* (2, 32)	7.13* (2, 52)

$*P < 0.05$

Declining species showed a trend towards having increasing Taylor's exponents (Fig. 2). This means that in declining species, high abundance populations at the center of the range became more variable relative to peripheral populations over time. Because of the interaction, increasing species showed the opposite trend so that high abundance populations became relatively less variable compared to peripheral populations (Fig. 2). Populations with higher temporal variability tend to be less stable than those with lower variability (Pimm 1991, Curnutt et al. 1996). This implies that species showing range-wide declines had centrally-located high abundance populations that became increasingly variable and hence less stable over time.

The short-term population trends analyzed here may be reversed over longer periods of time or show other kinds of range wide patterns of population change. However, our results imply that sustained declines or increases in species have clear consequences of geographic range structure. Thus, as biotic homogenization occurs throughout a large geographic region, changes in the relative stability of core populations and the degree to which the range becomes fragmented will follow. The geographic patterns that arise from biotic homogenization should emerge as follows. Human encroachments on natural habitats change them from habitats that once supported relatively stable and persistent populations of native species to habitats that cannot support large, stable populations across such species' the ranges. Native species decline in abundance and their ranges shrink to those few habitats that serve as refuges. The populations that remain, on average, are more widely separated than they previously were, so the geographic range becomes more fragmented. At the same time, the transformed habitats become suitable to sustain stable populations of weedy species, which then increase in abundance and spread throughout the human dominated landscape. As they spread, their geographic ranges become less fragmented. The macroecological data we present here shows that this process actually occurs over short time scales. Ultimately, such processes must underlie the replacement of native species adapted to unperturbed habitats with weedy species capable of surviving in human dominated landscapes.

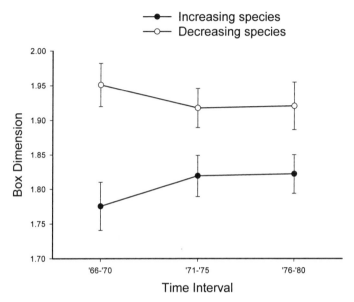

Figure 1. Changes in box dimensions of geographic ranges for 21 species demonstrating widespread population increases and 15 species showing widespread declines in North America. When range size was included as a covariate in the RMANOVA, there was a significant interaction between time and population trend, indicating increasing and decreasing species showed different sequences of change in fragmentation: increasing species' ranges became less fragmented, decreasing species ranges' became more fragmented. Note that the box dimensions shown here are not adjusted for differences in range sizes between species.

4. THE DYNAMICS OF GEOGRAPHIC RANGE EXPANSION BY WEEDY SPECIES

An important component of the homogenization process is the expansion of weedy species into geographic regions they formerly did not occupy. Some range expansions in ecological time are undoubtedly "natural" in the sense that they occurred largely in the absence of widespread changes induced by human activities. Those, however, that are the consequence of specific changes caused by urbanization, agriculture, and other human manipulations of ecosystems should provide a picture of how weedy species spread over time. In this section, we examine two specific invasions that are clearly due to human activities. This is justified by two general observations about these invasions. First, each of the two species considered here are clearly linked to human dominated ecosystems. Second, each of the species was introduced by human activities into urbanized ecosystems, and the subsequent spread of each of these species occurred most rapidly in human dominated landscapes. Both species have relatively good records of population dynamics associated with expansion of their geographic ranges, and these records are extremely useful in establishing some of the general properties associated with biotic invasions by weedy species.

European Starlings (*Sturnus vulgaris*) were introduced in the New York City area near the end of the last century (roughly 1890). There were several different introductions in the locality, and it was not immediately evident that the introduction would succeed. Wing (1941) tracked the initial progress of the invasion using data from the Christmas Bird Count (CBC) program. This program has been going on since the beginning of this century, and entails counts of the number of individuals of all species recorded within specific localities during December. Wing (1941) drew rough contours marking the progress of the invasion based on CBC data and other anecdotal observations made by individuals with whom he had correspondence. His approach was not quantitative, and he did not use any measures of relative abundance associated with the CBC.

From the geographic range paradigm, we initially would predict that the rate of spread of starlings should be most rapid through habitats that currently provide the best ecological conditions for the species. We cannot establish at this juncture what particular sets of conditions are best for starlings, however, it is evident that those conditions are closely associated with human dominated habitats because there is a strong positive relationship between human population abundance and starling abundance from counties in which BBS routes have been conducted (Fig. 3).

In order to describe the pattern of invasion of European starlings across the North American continent, we obtained data from the CBC, and used it to establish the location of the geographic range boundary of the invasion at five year intervals starting in 1905 and ending in 1960, when starlings had colonized the entire continent. We defined the range boundary as a convex polygon that connected the extreme points in the geographic distribution within each five-year period. We estimated relative population size by adding abundances on all CBC routes on which starlings were found for each five-year period and dividing the abundance in a period by the highest abundance (which occurred around 1975). Area of the geographic range was

8. A Geographic Perspective on Biotic Homogenization

estimated by calculating the area enclosed by the convex polygon describing the range boundary for each year since 1901.

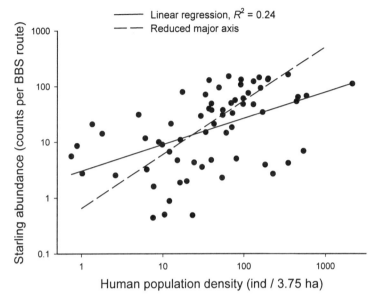

Figure 2. Changes in Taylor's exponents for 21 species demonstrating widespread population increases and 15 species showing widespread declines in North America. There was a significant interaction between time and population trend, indicating increasing and decreasing species showed different sequences of changes in the population stability of central populations: increasing species had central populations that became more stable, decreasing species had central populations that became less stable.

To estimate the rate of spread of starlings across the continent, we established several transects along the northern, southern, and western boundaries of the range connecting range boundaries from one five year period with the boundaries in the next period. We measured the length of the transect in km using ARCVIEW software extensions. We averaged the abundance at CBC sites along each transect during the five year period. From these data, we calculated the per capita rate of expansion of the geographic range (km y^{-1} indivdual^{-1}) by dividing the length of the transect by the product of the average abundance during the five year period and the number of years (five). We plotted the per capita rate of expansion of the range against current average BBS abundance along the transect. The

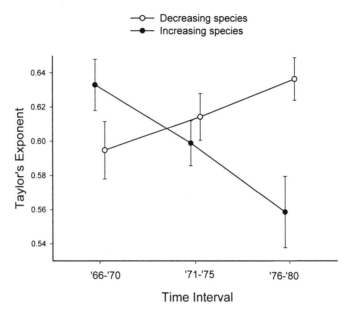

Figure 3. The relationship between current starling abundance and current human population density within 72 randomly selected US counties is positive. Since there is considerable uncertainty associated with estimates of human abundances, reduced major axis regression (RMA) was used to estimate the linear relationship (Sokal and Rohlf 1995).

average abundance from the BBS we used as an estimate of the suitability of the habitat through which the expansion occurred (Fig. 3).

The relative population size of European Starlings in North America remained quite low until about 1940, then expanded rapidly until 1975, after which it declined to about 60% of its maximum size (Fig. 4). The size of the area covered by the invasion also remained relatively small until about 1920, then expanded rapidly until 1935, after which the area continued to increase, but at a decreasing rate (Fig. 5). Note that population size lagged behind geographic range size, and only increased significantly after the invasion front began to slow as it approached the west coast of North America. From these patterns, we recognize several periods marking different phases of the invasion. From 1905 to 1920, there was slow range expansion, and slow population growth. Beginning in 1921 and proceeding through 1935, there was a rapid expansion of range size, but abundance remained at relatively low levels. Beginning in 1936, when the rate of range size growth was at its maximum, population abundance began to increase rapidly. By 1945, range size expansion was slowing significantly, but abundance was increasing at its maximum rate. After 1960, range size stopped growing but population abundance increased until about 1980, after which it dropped to current levels. When we plotted data from all transects across the entire 60 years of

8. A Geographic Perspective on Biotic Homogenization 167

the expansion of European Starlings in North America, we found no relationship between rate of expansion and current abundance (Fig. 6). This obscured the changes in the relationship between the per capita rate of expansion and habitat suitability, which changed over time. During the initial spread, from 1905 until 1920, there was no relationship. During this period of time, the species moved relatively slowly through New England and the Mid-Atlantic states. When the range began to increase rapidly after 1920, there was a positive relationship between rate of range expansion and habitat suitability (Fig. 7A). During this period of time the invasion passed through the Midwest and also towards the southwest into the southern United States. In both of these regions, starlings are currently quite abundant (Price et al. 1995:173). When the rate at which geographic range size was expanding began to decline after 1935, the per capita rate of expansion was negatively related to habitat suitability (Figure 7B). During this period of time, starlings were expanding rapidly across the Great Plains, Great Basin, and southwestern United States, regions where the species is currently less abundant (Price et al. 1995). After 1945, there was again no relationship between the per capita rate of spread and habitat quality as the species filled regions along the west coast of North America where it is relatively abundant in human dominated landscapes (Price et al. 1995).

The spread of European Starlings across North America followed four relatively distinct phases. During the first phase, population size increased only slightly and range expansion was relatively slow. During the second phase, the invasion proceeded rapidly through favorable habitats. The next phase was characterized by a slowing in the rate of expansion as the species encountered less favorable habitat. Much of the expansion during this time could have been due to the momentum gathered in the previous phase. Finally, the last phase was characterized by a slowing and eventual halt to the spread while the population within the range increased rapidly, then declined. The first three phases seem to correspond to patterns seen in a number of different species (Brown and Lomolino 1998:272-273). What has not been documented before, however, is that the major growth in abundance occurred *after* range size had ceased to increase. That is, it was only after individuals could disperse no further past existing range boundaries that population density within those boundaries increased significantly.

The spread of European Starlings occurred unevenly across the range as population dynamics within local sites interacted with a heterogenous environment. Specific population mechanisms such as negative density dependence and Allee effects may have been involved in the spread (van den Borsch et al. 1992, Veit and Lewis 1996), but these cannot be assumed to have been homogeneous during the invasion. Rather, population processes must have been different in different geographic regions so that the spread

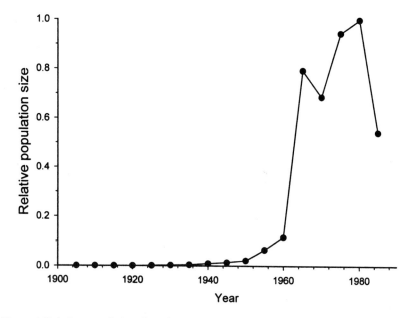

Figure 4. Relative population size of European Starlings from the date of their establishment in North America until 1985. Note that the temporal pattern is neither exponential nor logistic.

8. A Geographic Perspective on Biotic Homogenization 169

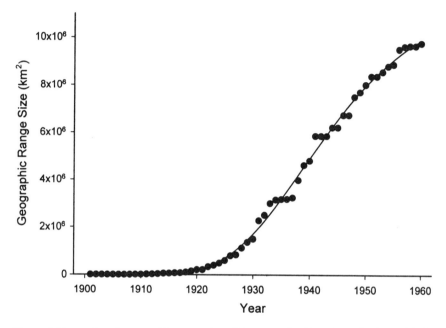

Figure 5. Dynamics of geographic range size of European Starlings from the time of their establishment in North America until 1960, when they had filled the entire continent. Note that the temporal pattern is approximately sigmoidal. The solid line represents a four parameter sigmoidal model of population change fit to the data.

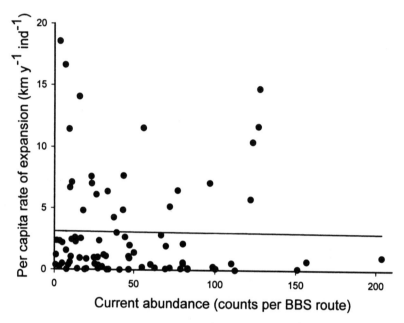

Figure 6. Per capita rates of range expansion plotted against current abundance for European Starlings obtained at different times during their invasion of North America. Current abundance is taken as a rough measure of habitat suitability. The solid line is from RMA regression. Note there is no relationship between rate of expansion and habitat suitability.

8. A Geographic Perspective on Biotic Homogenization

cannot be attributed solely to a particular population process: significant geographic variation in population dynamics must have caused these phases. This conclusion we suspect will apply to all invading species.

Unfortunately, the CBC are sufficiently coarse in their resolution that an adequate representation of the geographic pattern of local population dynamics that accompany an invasion are difficult to estimate. The recent invasion of the House Finch (*Carpodacus mexicanus*) through eastern North America, however, has been well documented in the BBS. Since records of House Finches were rare during the first 20 years of the BBS, we aggregated BBS data across two ten-year time periods (1966-1975, 1976-1985) for the first 20 years of the BBS. Expansion of the eastern range of the House finch was much more rapid after 1985 than before, so we aggregated BBS data over two five-year time periods (1986-1990, 1991-1994) subsequent to 1985. These data sets were used to examine the pattern of geographic variation in population dynamics that accompanied the spread of the species through eastern North America. We calculated the average per capita rate of population change for every route where House Finches were observed during each period from the beginning of the BBS (1966) until the end of the time period. This average was calculated as

$$r = \frac{\sum_{t=1}^{T}(\ln N_{t+1} - \ln N_t)}{T} \qquad \text{(EQUATION 4.1)}$$

where r is the per capita rate of change, N_t is the number of House Finches counted on a route at time t, and T is the period of time from 1966 until the end of the time period. We then plotted these estimates of per capita rates of change for each BBS route with House Finches on a map of eastern North America.

The general pattern that emerged from maps of average per capita rates of change was that the highest positive rates of change were found along the invasion front (Figure 8). Routes behind the invasion front appeared to either increase more slowly, show no growth, or decline slightly. Thus, the invasion of House Finches was typified by rapid population growth in local populations along the leading edge of the front followed by relative stability of those populations. At this time, since the invasion is still proceeding, and is now congealing with native populations of House Finches in western North America, it is not possible to examine the relationship of local population growth shortly after initial establishment of local populations with measures of habitat suitability. We predict, however, that there should be a positive relationship. That is, newly established populations should show the highest rates of initial population growth within those regions that prove to be the most suitable habitat.

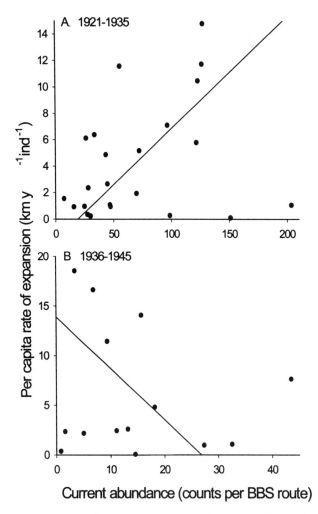

Figure 7. Per capita rates of range expansion plotted against current abundance for European Starlings obtained during different phases of their invasion of North America. Current abundance is taken as a rough measure of habitat suitability. Lines were obtained from RMA regression.

8. A Geographic Perspective on Biotic Homogenization

Figure 8. Estimates of average per capita rates of change of local populations (r) of House Finches during different time periods of their invasion of eastern North America. Solid triangles indicate r Æ 1, open triangles indicate 1 > r > 0, open circles indicate r œ 0. Top left figure 1966 - 1975. Top right figure 1976 - 1985. Bottom left figure 1986-1990. Bottom right figure 1991-1994.

5. THE GEOGRAPHY OF BIOTIC HOMOGENIZATION

The patterns we have documented here suggest that the large-scale homogenization of biotas that follows human changes in local ecosystems on a continent has a clear geographic signature. These patterns imply that a distinct sequence of ecological processes occur that eventually leads to wholesale biotic replacements within human dominated continental ecosystems. Obviously, the first part of this process is widespread alteration of ecosystems by human activities. Replacement of "natural" ecosystems that originally were spatially very diverse with more uniform agricultural or urban ecosystems homogenizes landscapes and the geographic regions containing them. When such alterations become spatially extensive, then geographic ranges of species that cannot exist in human dominated ecosystems become more fragmented and central, core populations become less stable. This increases the likelihood of both local and global extinction for such species. Extensive human-caused changes in ecosystems also favor the expansion of species that can use such ecosystems. The expansion of these weedy species begins relatively slowly (in ecological time) with slow total population growth and expansion of the area of the invasion. As the invasion progresses, the geographic population of the invader picks up momentum as it moves through human dominated landscapes. Ecological success of local populations along the invasion front translates into increased momentum of the invasion front and more rapid range expansion. Range expansion occurs more rapidly than expansion of the total population, because dispersing individuals find better habitat along the wave front than in populations behind it. When range expansion reaches habitat limits, then the wave front slows down, and populations behind the range boundary increase until they reach saturation.

The data we have presented here describes fluctuations in geographic populations over short time periods and the early stages of range expansion by weedy species. None of the species examined here are in the final stages of range collapse wherein the species is extinct throughout a large fraction of the historical geographic range. Our data imply that the initial stages of range collapse involve destabilization of populations in the center of the geographic range coupled with increased fragmentation of populations across the range. Lomolino and Channel (1995, 1998; Channel and Lomolino 1999a) showed that species with drastic reductions of their historical ranges tended to persist in populations near the periphery of their geographic ranges. Channel and Lomolino (1999b) analyzed data on patterns of range reduction in species, and found evidence for a process where demographic declines were experienced in a non-uniform spatial pattern across the range. They called this pattern of decline "contagion" to suggest that population declines in endangered species occurred as caused by a set of factors that move from one side of the range to the other in relatively small jumps. Populations that remain within the range are those that for one reason or the other were isolated from the spreading pattern of local extinction. Such populations tend to be at the boundary of a species' range.

If we take our data as indications of initial stages of declines, we hypothesize that species with destabilized central populations and an increased degree of range fragmentation ought to be more susceptible to the kind of "epizootic-like" spread of local extinction that marks the trajectories of species that are nearing the brink of extinction due to human imposed ecological changes.

The process of biotic homogenization of a geographic region integrates ecological processes with geographic variation in environmental conditions to produce a relatively complete accounting of the changes in geographic ranges that accompany homogenization. It is at the geographic scale that the consequences of homogenization are likely to be most severe. If homogenization proceeds sufficiently rapidly and on a large enough geographic scale, the inevitable consequence will be widespread extinctions of species not adapted to human ecosystems. The evidence that we have accumulated here indicates that such consequences are indeed likely to occur. Whether the homogenization process has proceeded far enough up to this point to make widespread extinctions inevitable is not clear. Note, though, the significant lag time in population and geographic range expansion inherent in the invasions we describe here. It is not inconsistent to think that we are still in a phase where many weedy species have not yet overcome the initial inertia to widespread range expansion. However, there is the likelihood that many species are just now beginning their rapid phase of their range extensions. As more and more invasive species become established, their ecological influence may further erode the ecological suitability of habitats for native species that are still able to persist in geographic regions that are less impacted by human activities.

There is nothing qualitatively different between the geographical population processes that we have described here and those that must have occurred prehistorically as changes in global climates led to changes in prehistoric continental (or marine) ecosystems. There are many instances of mass extinctions that over the course of a few million years greatly simplified and homogenized biotas (Erwin 1998). Many changes in geographic ranges that led to extinctions of some species and establishment of others over evolutionary time must have occurred on roughly the same spatial scales that we have described here. The sole difference between such changes over evolutionary time and those occurring in a human dominated biosphere is that a single species is responsible for the changes in our current biosphere. These changes are thus occurring on a time scale relevant to human population dynamics, rather than on the time scale typical of most global changes (even though many such changes may be relatively rapid in geological time). This means that the time scale of current widespread global change is probably much more rapid than any widespread changes experienced by the biosphere in the past, since it is being driven by a species undergoing a rapid population expansion in ecological time. Thus, the changes humans are current imposing on the biosphere may be so quantitatively different from other such changes that we can look for only limited guidance to the past to help us understand the biosphere's current trajectory.

REFERENCES

Bright, C. 1996. Understanding the threat of bioinvasions. In: State of the world, 1996 (L. Starke, ed.), W.W. Norton, New York, Pp: 95-113.

Brown, J.H. 1984. On the relationship between distribution and abundance. American Naturalist 124:255-279.

Brown, J.H. 1995. Macroecology. University of Chicago Press, Chicago.

Brown, J.H. and B.A. Maurer. 1987. Evolution of species assemblages: effects of energetic constraints and species dynamics on the diversification of the North American terrestrial avifauna. American Naturalist 130:1-17.

Brown, J.H. and B.A. Maurer. 1989. Macroecology: the division of food and space among species on continents. Science 243:1145-1150.

Brown, J.H., D.W. Mehlman, and G.C. Stevens. 1995. Spatial variation in abundance. Ecology 76:2028-2043.

Channell, R., and M.V. Lomolino. 1999a. A geography of extinction: dynamic biogeography and conservation of endangered species. Nature, in press.

Channel, R., and M.V. Lomolino. 1999b. Trajectories to extinction: spatial dynamics of the contraction of geographic ranges. J. Biogeography, in press.

Cronk, Q.C.B., and J.L. Fuller. 1995. Plant invaders: the threat to natural ecosystems. Chapman and Hall, London.

Curnutt, J.C., S.L. Pimm, and B.A. Maurer. 1996. Population variability of sparrows in space and time. Oikos 76:131-144.

Erwin, D.H. The end and the beginning: recoveries from mass extinctions. Trends in Ecology and Evolution 13:344-349.

Hengeveld, R. 1989. Dynamics of biological invasions. Chapman and Hall, London.

Hengeveld, R., and J. Haeck. 1981. The distribution of abundance: II. Models and implications. Proc. Konink Nederlandse Akademie van Wetenshappen, series C 84:257-284.

Hengeveld, R., and J. Haeck. 1982. The distribution of abundance: I. Measurements. J. Biogeography 9:303-316.
James, F. C., C.E. McCulloch, and D.A. Wiedenfeld. 1996. New approaches to the analysis of population trends in land birds. Ecology 77:13-27.
Linder, E.T., M.-A. Villard, B.A. Maurer, and E.V. Schmidt. In press. Geographic range fragmentation and abundance in North American landbirds: variation with migratory status, trophic level, and breeding habitat. Ecography
Lomolino, M.V., and R. Channel. 1995. Splendid isolation: patterns of range collapse in endangered mammals. J. Mammalogy 76:335-347.
Lomolino, M.V., and R. Channel. 1998. Range collapse, reintroductions and biogeographic guideline for conservation: a cautionary note. Conservation Biology 12:481-484.
Maurer, B.A. 1994. Geographical population analysis. Blackwell Scientific, Oxford.
Maurer, B.A. 1999. Untangling ecological complexity. University of Chicago Press, Chicago.
Maurer, B.A., and J.H. Brown. 1989. Distributional consequences of spatial variation in local demographic processes. Annales Zoologici Fennici 26:121-131.
Maurer, B. A. and M. P. Nott. 1998. Geographic range fragmentation and the evolution of biological diversity. In: Biodiversity dynamics: turnover of populations, taxa, and communities (McKinney, M. L. and J.A. Drake, eds.), Columbia University Press, New York.
Maurer, B.A. and M.A. Villard. 1994. Geographic variation in abundance of North American birds. National Geographic Research and Exploration 10:306-317.
Pimm, S.L. 1991. Balance of nature? University of Chicago Press, Chicago.
Price, J., S. Droege, and A. Price. 1995. The summer atlas of North American birds. Academic Press, Inc., San Diego.
Robbins, C.S., D. Bystrak and P.H. Geissler. 1986. The Breeding Bird Survey: its first fifteen years. Resource Publication 157, US Dept Interior, Fish Wildlife Service, Washington DC.
Sauer, J. R., J. E. Hines, G. Gough, I. Thomas, and B. G. Peterjohn. 1997. The North American Breeding Bird Survey Results and Analysis. Version 96.4. Patuxent Wildlife Research Center, Laurel, MD.
Sokal, R., and F.J. Rohlf. 1995. Biometry, 3rd ed. Freeman, New York.
Skellam, J.G. 1951. Random dispersal in theoretical populations. Biometrika 38:196-218.
Taylor, L.R. 1961. Aggregation, variance, and the mean. Nature 189:732-735.
Taylor, L.R. and R.A.J. Taylor. 1977. Aggregation, migration, and population mechanics. Nature 265:415-421.
Turner, B.L., C. Clark, R.W. Kates, J.F. Richards, J.T. Mathews, and W.B. Meyer. 1990. The earth as transformed by human action. Cambridge University Press, Cambridge.
van den Bosch, F., R. Hengeveld, and J.A.J. Metz. 1992. Analysing the velocity of animal range expansion. Journal of Biogeography 19:135-150.
Veit, R.R., and M.A. Lewis. 1996. Dispersal, population growth, and the Allee effect: dynamics of the House Finch invasion of eastern North America. American Naturalist 148:255-274.
Villard, M.-A., and B.A. Maurer. 1996. Geostatistics as a tool for examining hypothesized declines in migratory songbirds. Ecology 77:59-68.
Williams, J.D., and G.K. Meffe. 1999. Nonindigenous species. In: Status and trends of the nation's biological resources (Mac, M.J., P.A. Opler, C.E. Puckett Haecker, and P.D. Doran, eds.), U.S. Department of the Interior, U.S. Geological Survey, Reston, Virginia, Pp: 117-129
Williamson, M.H. 1996. Biological invasions. Chapman and Hall, London.
Wing, L. 1941. Spread of the Starling and English Sparrow. Auk 60:74-87.

Chapter 9

Global Warming, Temperature Homogenization and Species Extinction

J.L. Green[1], J. Harte[2,3], A. Ostling[2]
[1]*Department of Nuclear Engineering, University of California, Berkeley, CA 94720, USA*
[2]*Energy and Resources Group, University of California, Berkeley, CA 94720, USA*
[3]*Department of Environmental Science, Policy, and Management, University of California, Berkeley, CA 94720, USA*

1. INTRODUCTION

The global mean surface air temperature has increased by about 0.5 °C over the past century, and is expected to increase by 1.5 to 4.0 °C in response to a doubling of atmospheric CO_2 (relative to the pre-industrial level) over the next century (IPCC 1996). Variations in the rise in temperature at regional to continental scales are expected to depend on both latitude and altitude. General circulation models (GCMs) project a maximum mean warming in high latitudes, in part a consequence of ice-albedo feedback associated with reduced sea ice cover (IPCC 1996). At high elevations, a similar feedback-induced temperature enhancement is likely to occur if warming reduces the duration of snowcover. Hence, the equator-to-pole and sea-level-to-mountaintop temperature gradients will likely shrink. This predicted climate change would homogenize the temperature of the earth; the ecological effect of this abiotic homogenization is the focus of this chapter.

Climate plays a major role in shaping the geographical distributions of species. As climate changes, it is expected that many species' ranges will shift, shrink, or disappear entirely (IPCC 1996; Peters 1992). Small temperature changes of less than one degree within this century have been associated with substantial range changes for both butterflies and birds. Parmesan (1996) documented a shift in the range of the Ediths checkerspot butterfly (*Euphydyras editha*) in western North America. Moreover, Parmesan et al. (1999) documented a poleward shift for a group of species' ranges. In a sample of 35 non-migratory European butterflies, the ranges of 22 species shifted to the north by 35 – 240 km. Two-thirds of these 22 species expanded their range when shifting northward, while the remaining species shifted their range northward without altering the latitudinal extent of their range. Thomas and Lennon (1999) documented range shifts over the past century in Europe for two groups of breeding birds: those restricted to the south of Britain and those restricted to the north. They found an average 18.9 km shift in the northern boundary of 59 southerly bird species, and no consistent shift in the southern boundary of 42 northerly bird species. These results parallel the results obtained by Parmesan et al. (1999) for European butterflies, where the northern margins have expanded more than the southern margins have retracted.

A simplified explanation of the aforementioned range shifts in European butterflies and birds is that rising temperatures in Europe altered the ratio of extinctions to colonizations at the northern and southern boundaries of the species' ranges. As temperatures rose, species either experienced increased colonization of new habitats toward the pole, or increased extinctions at the equatorial range boundary due to unsuitable conditions, or both. The resulting combination of net extinctions at the low latitude range boundary and/or net colonizations at the high latitude range boundary caused an overall poleward range shift (Parmesan et al. 1999).

Response to climate change over the next century will be species dependent. Some migratory species may be able to alter the timing or destination of migration in response to annual temperature variations. Mobile species in general can experience population level distribution changes via colonization and extinction at range boundaries, as described above for butterflies and breeding birds. Sessile species (e.g., plants) on the other hand, may not be able to colonize poleward rapidly enough to move away from potentially intolerable temperature extremes imposed by climate change. Sessile individuals subjected to intolerable temperatures will perish or fail to reproduce, and thus cause species ranges to contract.

It is predicted that as climate warms, suitable habitat for species will shift upward in altitude as well as shifting poleward in latitude (Peters and Darling 1985, McDonald and Brown 1992, IPCC 1996). Generally, a short rise in elevation corresponds to a large shift in latitude: a 3 °C cooling due to a 500 meter rise in elevation on a moist mountain is roughly equivalent to a 300 kilometer poleward shift in latitude (MacArthur 1972). Because mountains shrink in area with elevation, species shifting upward in response to warming will typically occupy smaller areas, have smaller populations, and thus become more susceptible to environmental stress and genetic pressure. Species originally situated near mountain summits, with no habitat to move up to, may be entirely displaced by the species moving in from lower elevations (Peters 1992). Using data on the distributions of boreal mammals inhabiting montane forests of the Great Basin, McDonald and Brown (1992) predicted how many and which species will be lost from these forests in response to global warming. For a 3 °C temperature rise, they predicted 45 extirpations across nineteen mountain ranges, with four of those ranges losing at least half of their species (Brown and Lomolino 1998).

Species that experience latitudinal range shifts due to global warming may experience either range contraction or expansion, depending on land mass shape and climate change scenario. If, for example, the northern and southern boundaries of species' ranges shift poleward in latitude on a continental land mass which shrinks with latitude (e.g., South America or Africa), these species will be forced to inhabit a smaller area. Species situated at the poleward margin of their range might have no suitable habitat to colonize at higher latitudes, and will thus either perish due to warming or face increased competition with species moving in from lower latitudes. In contrast to the aforementioned example of range contraction, species' ranges might expand in scenarios where warming is magnified with latitude.

The dependence of species richness on area has been well documented and the general patterns are widely established in ecology (Rosenzweig 1995). In this chapter, we explore the interplay between global warming, species range shifts, changes in inhabited area, and species richness. Using a species-area relationship and an endemics-area relationship, we develop quantitative insights into how the shape of landmasses combines with patterns of warming, such as temperature homogenization, to affect the survival of both mobile and sessile species. Our methods are general and may be applied to a wide range of biome types and climate warming scenarios.

2. SPECIES-AREA RELATIONSHIPS

To assess species loss due to global warming, we use two relationships between area and the number of species contained in that area. First, we apply a power-law form of the species-area relationship (SAR), which has been empirically tested in many systems over a large range of scales (Rosenzweig 1995). Second, we will apply a related endemics-area relationship, which has been shown to follow from the SAR (Harte and Kinzig 1997). In this section we summarize the nature of these two relationships.

According to the power-law form of the SAR, the number of species within broadly specified taxonomic groups (such as birds or plants) scales with area as:

$$S(A) = S_0 \left(\frac{A}{A_0}\right)^z \qquad \text{(EQUATION 2.1)}$$

or, equivalently

$$S(A) = cA^z \qquad \text{(EQUATION 2.2)}$$

where $S(A)$ is the total number of species in a censused patch of area A, S_o is the number of species found in an entire biome of area A_o, and z is a constant called the SAR exponent which is bounded between 0 and 1.

Harte and Kinzig (1997) and Harte et al. (1999a) have demonstrated that the power law form of the SAR is analytically equivalent to a condition of self-similarity in the spatial distribution of species. Several ecological relationships arise from this self-similarity condition, one of which is an endemics-area relationship (EAR). The EAR states that the number of species confined to smaller patches within a larger biome scales with area according to a power-law relationship:

$$E(A) = S_0 \left(\frac{A}{A_0}\right)^{z'} \qquad \text{(EQUATION 2.3)}$$

or, equivalently

$$E(A) = c'A^{z'} \qquad \text{(EQUATION 2.4)}$$

where $E(A)$ is the number of species found *only* in an area A and nowhere else in a larger biome of area A_o, and z' is a constant related to the SAR exponent by the formula

$$2^{-z'} = 1 - 2^{-z}. \qquad \text{(EQUATION 2.5)}$$

The constraint $z \leq 1$ implies $z' \geq 1$, and for $z = 0.25$ (a typical value), $z' = 2.65$. Note that when using the EAR, A_o must be large enough such that most of the S_o species found in A_o are endemic to A_o for the taxon of interest.

It has been demonstrated both empirically and theoretically (Kunin 1997; Harte et al. 1999b; Harte and Kinzig 1997) that species richness depends on patch shape (for example, the length to width ratio of a rectangular patch). It has been observed that long, skinny rectangular patches contain more species than square patches of the same area. Strictly speaking, the SAR and EAR results from self-similarity only hold for rectangles, with a length-to-width

9. Global Warming, Temperature Homogenization, and Extinction

ratio of √2; these rectangles have the property that successive bisections are shape-preserving (Harte et al. 1999c). The more a patch deviates from this shape, the less accurate are the SAR and the EAR for those patches. We refer below to patches as being "well shaped" if their length to width ratio is √2.

In several of the following examples, we estimate species loss following climate change for a biome with length to width ratio 2:1. The most accurate method of estimation in these examples involves applying either the SAR or the EAR to smaller patches within the biome obtained from successive bisections that maintain a length to width ratio of 2:1, or as similar to that shape as possible. Harte et al. (1999b) and Kinzig and Harte (in press) describe methods of analysis for odd shaped patches.

3. THE MODEL

The dependence of species' geographical distribution on temperature, the ability of species to respond to climate change, the shape of areas in which species live, and the temperature rise across latitude all have many complicated nuances. In order to proceed with an analysis of species loss due to climate change, we create a model in which these characteristics of the problem are confined to a simplified set of possibilities. In the Discussion section we explore the consequences of these simplifications. Our model consists of the following methods and assumptions:

1. We examine two biome shapes: rectangles and triangles (with the apex pointed poleward).
2. Species fall into two discrete categories: totally mobile, or totally immobile. Species that have the ability to shift their range in response to climate change are considered mobile. Species that are unable to shift their range we refer to as sessile.
3. A species' range depends on land surface temperature and the location of its biome. A species endemic to a particular range is confined by the temperature conditions of that range and the boundaries of its biome.
4. Following climate change, mobile species will shift their range boundaries to track their climatic tolerance. Therefore, mobile species that inhabit a range between two specific continental isotherms will relocate in response to climate change such that they remain between these two isotherms.
5. Sessile species are not capable of shifting their range boundaries to track their climatic tolerance. Following climate change, sessile species that once inhabited a range between two specific continental isotherms will be subjected to a new pair of hotter continental isotherms at the northern and

southern boundaries of their range. If the entire range following climate change is hotter than the warmer of the two bordering isotherms before the climate change, the sessile species endemic to this range will go extinct.

6. Neither sessile nor mobile species are capable of survival outside of their original (pre-warming) biome (which we denote as A_o). Species ranges may be smaller than the area encompassed by the biome, but never larger. Therefore, following global warming, if a species' range is shifted entirely outside of the biome, that species will go extinct*.

7. Pre-warming land-surface temperature decreases linearly with increasing latitude. This implies that isotherms run parallel to latitude lines and are equally spaced.

8. We look at two warming scenarios: a temperature rise that is uniform within a biome across latitude lines, and a temperature rise that increases linearly with latitude. A rise in temperature that increases linearly with latitude will homogenize the temperature of the earth; we call this type of warming scenario "homogenizing". A temperature rise that is uniform across a biome we call "non-homogenizing".

A consequence of 7 is a direct relationship between both the magnitude and distribution of the temperature rise and the magnitude and distribution of the isotherm shift, and hence the range shift (Figure 1). A uniform temperature rise across a biome will lead to a uniform range shift. A temperature rise that increases linearly with latitude will lead to a range shift that increases linearly with latitude, such that ranges in the poleward region of a biome experience larger latitudinal shifts than those nearer the equator. In accordance with our definitions in 8, we call each of these range shifts non-homogenizing and homogenizing, respectively.

3.1   Examples

The following examples are intended to illustrate how we apply the SAR and EAR to calculate the fraction of species lost within a biome following climate change, in the context of our simplified model. To demonstrate our approach, we present several different examples with varying combinations of biome shape, species type and climate change scenario.

Although by our assumptions species' ranges are confined by the boundaries of the biome, our conceptual process involves allowing species' ranges to shift beyond the biome boundary following global warming such that they track their climatic tolerance. Once we have established these post-warming ranges, we use the assumption that species cannot survive outside their original (pre-warming) biome to estimate species loss.

9. Global Warming, Temperature Homogenization, and Extinction

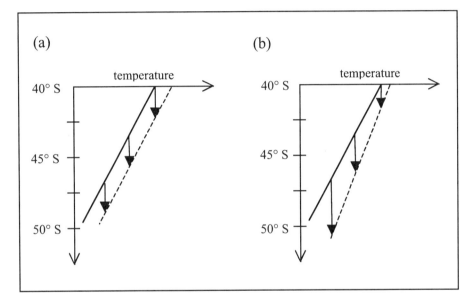

Figure 1. Examples of climate change and species' range shifts in the southern hemisphere. Solid and dashed lines represent temperature gradients pre- and post-warming, respectively. Arrows reflect the magnitude of range shifts. (a) Non-homogenizing temperature rise resulting in a range shift of ~ 2.5° latitude. (b) Homogenizing temperature rise and range shift. Note that both the temperature rise and range shift increase linearly with latitude.

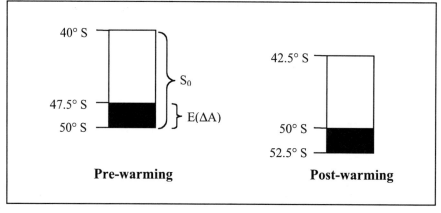

Figure 2. Example of a mobile species range shift following a non-homogenizing temperature rise. Before the climate change, S_o species inhabit a biome between 40° S and 50° S. Following the temperature rise, all S_o species experience a poleward non-homogeneous range shift of 2.5° latitude. Species once endemic to the area ΔA between 47.5° S and 50° S are forced out of the biome, and thus go extinct.

9. Global Warming, Temperature Homogenization, and Extinction

Consider a rectangular biome that exists precisely between 40° S and 50° S, with length to width ratio 2:1, harboring S_o mobile species of a particular taxonomic group in an area A_o (Figure 2). Assume that this biome has been subjected to a non-homogenizing temperature rise, resulting in a species range shift of exactly 2.5° in latitude. According to our assumptions, the ranges of mobile species in the biome will move poleward 2.5° in latitude (i.e., a ~ 3°C warming), and all mobile species whose range is contained entirely between 47.5° S and 50° S before the temperature change will go extinct.

We employ two different methods for calculating the fraction of mobile species lost within a biome following a change in climate. The first method is to use Equation 2.3 to count up the species lost, and divide by the total number of species originally in the biome:

$$\frac{E(\Delta A)}{S_0} = \left(\frac{\Delta A}{A_0}\right)^{z'}, \qquad \text{(EQUATION 3.1.1)}$$

where ΔA is the portion of the biome's climatic range which shifted out of the biome's original area, and hence $E(\Delta A)$ is the number of mobile species whose range is shifted entirely out of the biome. We call this method the EAR approach.

The second method for evaluating the fraction of mobile species lost is to use Equation 2.1 to calculate the total number of surviving species, and divide by the number of species originally in the biome:

$$\frac{S_0 - S(A_0 - \Delta A)}{S_0} = 1 - \left(\frac{A_0 - \Delta A}{A_0}\right)^z. \qquad \text{(EQUATION 3.1.2)}$$

$S(A_o-\Delta A)$ is the number of mobile species whose range did not entirely shift out of the biome. We call this method the SAR approach.

When both ΔA and $A_o - \Delta A$ are "well-shaped", these two methods yield the same result. For this example, ΔA is the region between 47.5 °S and 50 °S. This area has the same length to width ratio as the biome (2:1) and can be obtained by bisecting A_o twice. The area between 40 °S and 47.5 °S ($A_o - \Delta A$), on the other hand, has a 1_:1 length to width ratio and cannot be obtained from A_o by shape-preserving bisections. Although neither ΔA nor $A_o - \Delta A$ is "well shaped", to calculate the fraction of mobile species lost for this example, we choose the method which can be applied to the patch most similar in shape to A_o, i.e. ΔA. Applying the EAR approach to ΔA, the fraction of species lost following the non-homogenizing temperature change is predicted by:

$$\frac{E(\Delta A)}{S_0} = \left(\frac{\Delta A}{A_0}\right)^{z'} = \left(\frac{2.5w}{10w}\right)^{z'} = \left(\frac{2.5}{10}\right)^{z'}, \qquad \text{(EQUATION 3.1.3)}$$

where w is the width of the biome. If $z = .25$, then $z' = 2.65$ and the fraction of mobile species lost in the biome is approximately 3%.

Using the same rectangular biome and non-homogenizing temperature change, we now consider the fate of sessile species. Consider first the species endemic to the habitat bound between 40° S and 42.5° S. If these species were mobile, they would shift their range poleward 2.5 degrees in latitude. But sessile species are incapable of shifting their ranges to track their climatic tolerance. Therefore, all sessile species endemic to the habitat bound between 40° S and 42.5° S will go extinct. Likewise, species endemic to the habitat bound between 40.5 ° S and 43° S will go extinct. Yet, if we attempted to calculate the total number of sessile species lost between 40° S and 43° S by summing those endemic to 40° S – 42.5° S with those endemic to 40.5° S – 43° S, we would double count the species endemic to 40.5° S – 42.5° S. Figure 3 illustrates the logic for predicting the total number of sessile species lost in the biome between 40° S and 50° S: we "step down" the biome in discrete latitudinal increments, summing up the endemics lost within 2.5° latitude bands and subtracting those endemics previously accounted for.

If we "step down" the biome in arbitrarily chosen 0.5° latitude increments, the sum of endemics lost within each 2.5° latitude band minus the endemics previously accounted can be expressed as:

$$E(A_{40°S-42.5°S}) + E(A_{40.5°S-43°S}) - E(A_{40.5°S-42.5°S})$$
$$+ E(A_{41°S-43.5°S}) - E(A_{41°S-43°S}) + \ldots$$
$$+ E(A_{47.5°S-50°S}) - E(A_{47.5°S-49.5°S})$$

$$= E(\Delta A) + 15[E(\Delta A) - E(\Delta A - w\delta)]$$

(EQUATION 3.1.4)

where ΔA and $w\delta$ correspond to the area of a 2.5° and 0.5° latitudinal band within the biome, respectively.

Using the EAR method and Equation 3.1.4, the predicted fraction of sessile species lost in the biome is:

9. Global Warming, Temperature Homogenization, and Extinction

$$\frac{E(\Delta A) + 15[E(\Delta A) - E(\Delta A - w\delta)]}{S_0} = 16\left(\frac{2.5}{10}\right)^{z'} - 15\left(\frac{2}{10}\right)^{z'}$$

(EQUATION 3.1.5)

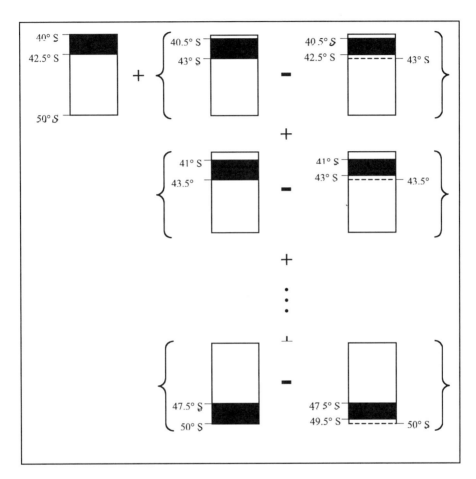

Figure 3. This figure illustrates the process for calculating sessile species loss following a non-homogenizing temperature change. For this particular change in climate, sessile species endemic to 2.5° latitudinal bands are subjected to intolerable temperature conditions. To calculate the fraction of species lost, we "step down" the biome in arbitrarily chosen discrete increments of 0.5° latitude, summing up the species endemic to each 2.5° latitude band and subtracting those which have been previously accounted for. The total number of sessile species lost is: $E(A_{40°S-42.5°S}) + E(A_{40.5°S-43°S}) - E(A_{40.5°S-42.5°S}) + E(A_{41°S-43.5°S}) - E(A_{41°S-43°S}) + \ldots + E(A_{47.5°S-50°S}) - E(A_{47.5°S-49.5°S})$.

Again, for $z = 0.25$ and $z' = 2.65$, the fraction of species lost is approximately 20%.

To approach the problem precisely, we "step down" the biome in increments of δ degrees, and then take the limit as $\delta \to 0$:

$$\lim_{\delta \to 0} \frac{E(\Delta A) + \frac{7.5}{\delta}[E(\Delta A) - E(\Delta A - w\delta)]}{S_0} =$$

$$\lim_{\delta \to 0} \left(1 + \frac{7.5}{\delta}\right)\left(\frac{2.5}{10}\right)^{z'} - \frac{7.5}{\delta}\left(\frac{2.5 - \delta}{10}\right)^{z'} = \left(\frac{2.5}{10}\right)^{z'} + \left(\frac{7.5z'}{10^{z'}}\right)(2.5)^{z'-1}$$

(EQUATION 3.1.6)

For $z' = 2.65$, the fraction of sessile species lost is approximately 23%.

Note that the application of the SAR for sessile species loss in this example would be much more complicated, due to the fact that there is not one contiguous area from which to calculate the fraction of species remaining in the biome. In other words, the area $A_o - \Delta A$ used in the SAR method (Equation 3.1.2) would be the sum of two different areas, physically separated by the area ΔA (i.e., the area ΔA used in Equations 3.1.4 – 3.1.6). Because the SAR method requires using one contiguous area, we have choosen the EAR method for predicting sessile species losses.

Next, we analyze a scenario in which the temperature rise is homogenizing; i.e., both temperature rise and range shift increase linearly with latitude. Specifically, we will consider the fraction of mobile species lost from the rectangular biome used in the examples above, with a range shift that increases linearly from 1.5° latitude at the equatorial facing edge to 3.5° latitude at the poleward facing edge. Note that this implies an average temperature change which is equivalent to that used in the previous examples.

Mobile species subjected to a homogenizing temperature rise will not merely shift their entire range poleward. For the likely scenario in which the temperature rise increases towards the poles, species' ranges expand (Figure 4). Using the SAR, the fraction of species lost is:

$$\frac{S_0 - S(A_0 - \Delta A)}{S_0} = 1 - \left(\frac{A_0 - \Delta A}{A_0}\right)^z = 1 - \left(\frac{7}{10}\right)^z \quad \text{(EQUATION 3.1.7)}$$

9. Global Warming, Temperature Homogenization, and Extinction

which for $z = 0.25$ is approximately 8%. Using the EAR, the fraction of species lost is:

$$\frac{E(\Delta A)}{S_0} = \left(\frac{\Delta A}{A_0}\right)^{z'} = \left(\frac{3}{10}\right)^{z'}, \qquad \text{(EQUATION 3.1.8)}$$

which for $z' = 2.65$ is approximately 4%. Because neither of the patches A and ΔA are "well shaped" or similar in shape to A_o, we use both the SAR and EAR method as a best estimate, and therefore predict a range in the fraction of species lost between 4% and 8%.

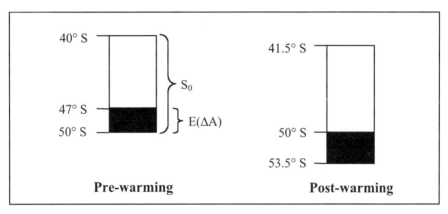

Figure 4. Mobile species range shift following a homogenizing temperature rise. Before the climate change, S_o species inhabit a biome between 40° S and 50° S. Following the temperature rise, the S_o species experience a poleward homogenizing range shift which varies linearly from 1.5° latitude at the northern edge of the biome to 3.5° latitude at the southern edge of the biome. Species once endemic to the area ΔA between 47° S and 50° S have their range shifted entirely out of the biome, and thus go extinct.

As a final example, we explore what happens for a non-homogenizing temperature change, for a biome in which the latitudinal dimension shrinks with latitude. Consider the hypothetical triangular biome with a base one-half the length in height, situated between 40° S and 50° S as illustrated in Figure 5. We can first estimate the fraction of species lost using the species endemic to the tip of the triangle, which are forced out of the biome. Using the EAR method, the fraction of species lost is:

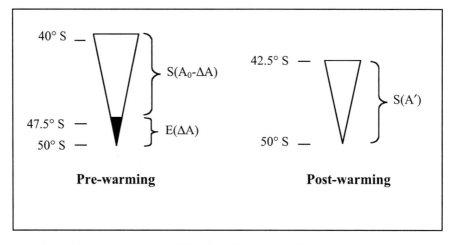

Figure 5. Mobile species range shift of 2.5° latitude following a non-homogenizing temperature rise for a biome in the shape of a triangle. $S(A_0-\Delta A)$ is the total number of mobile species in the trapezoid $A_0-\Delta A$, which shift into the trapezoid A' following climate change. Due to crowding effects, only $S(A')$ of these species survive at equilibrium. The $E(\Delta A)$ mobile species endemic to ΔA between 47° S and 50° S have their range shifted entirely out of the biome, and thus go extinct.

9. Global Warming, Temperature Homogenization, and Extinction 193

$$\frac{E(\Delta A)}{S_0} = \left(\frac{\Delta A}{A_0}\right)^{z'} = \left(\frac{\frac{1}{2}(1.25)(2.5)}{\frac{1}{2}(5)(10)}\right)^{z'}, \qquad \text{(EQUATION 3.1.9)}$$

which for $z' = 2.65$ is $\sim .1\%$.

Next, we can estimate the fraction of species lost by assuming that the species which reside in the trapezoid between 40° S – 47.5° S, are all capable of relocating into the triangle encompassed between 42.5° S – 50° S. This essentially means that all species forced to move from the trapezoid of area A_0-ΔA, into a triangle of smaller area A', are able to survive under crowded conditions. Using the SAR method for this case, the fraction of species lost is:

$$\frac{S_0 - S(A_0 - \Delta A)}{S_0} = 1 - \left(\frac{A_0 - \Delta A}{A_0}\right)^z = 1 - \left(\frac{\frac{7.5}{2}(5+1.25)}{\frac{1}{2}(5)(10)}\right)^z,$$

(EQUATION 3.1.10)

which for $z = .25$ is $\sim 2\%$.

Over time the crowding of species will lead to further extinction, however. The number of species lost when equilibrium is established can be estimated by applying the SAR method to the remaining habitable area within the biome after warming, which is represented by the smaller triangle between 42.5° S and 50° S of area A'. The fraction of species lost is thus:

$$\frac{S_0 - S(A')}{S_0} = 1 - \left(\frac{A'}{A_0}\right)^z = 1 - \left(\frac{\frac{1}{2}(3.75)(7.5)}{\frac{1}{2}(5)(10)}\right)^z,$$

(EQUATION 3.1.11)

which for $z = .25$ equals 13%.

Note that for the all three methods used in this example, we calculated the fraction of species lost using the same power-law exponent z. We therefore assumed that under both crowded and equilibrium conditions, species richness is governed by the same pre-warming SAR exponent z. Furthermore, none of the shapes used to calculate the SAR and EAR in this example are "well shaped" rectangles. The calculated losses in species richness are thus our best available estimates. Due to factors such as species competition under crowded conditions, the most realistic estimate for the fraction of species loss is under equilibrium conditions. Hence, we estimate the fraction of species lost as 13%.

4. DISCUSSION

4.1 Trends

In addition to the examples given in Section 3, we have calculated the fraction of species lost for three other scenarios: 1) sessile species loss for a rectangular biome subjected to a homogenizing temperature rise, 2) mobile species loss for a triangular biome subjected to a homogenizing temperature rise, 3) sessile species loss in a triangular biome subjected to a non-homogenizing temperature rise. In order to keep our results comparable with the examples already demonstrated, we used the same biome dimensions for both rectangles and triangles, the same homogenizing and non-homogenizing temperature changes, and the same species-area exponent $z = .25$. Table 1 summarizes the calculated fraction of species lost for all scenarios that we have analyzed.

Table 1. Summary of the estimated fraction of species lost for different combinations of biome shape (triangle vs. rectangle), species mobility (mobile vs. sessile) and climate change scenario (non-homogenizing temperature rise vs. homogenizing temperature rise). The scenarios are listed in order of increasing fraction of species lost within the biome.

Biome Shape	Species Mobility		Warming		Estimated Fraction of Species Lost
	Mobile	Sessile	Non-Homog.	Homog.	
▭	X		X		3%
▭	X			X	4% - 8%
▽	X			X	8%
▽	X		X		13%
▭		X		X	19%
▭		X	X		23%
▽		X	X		42%

Table 2. Summary of trends in the fraction of species lost. Inequalities (> or ≥) denote which biome shape, species type, climate change scenario or SAR exponent z yield the greater fraction of species lost, in accordance with the variety of scenarios discussed in the text. Note that the climate change scenario with the greater fraction of species lost depends on both species mobility and biome shape.

Biome Shape	Triangle	≥	Rectangle
Species Mobility	Sessile	>	Mobile
Climate Change Scenario			
Mobile Species & Rectangular Biome	Homogenizing	>	Non-homogenizing
Mobile Species & Triangular Biome	Non-homogenizing	>	Homogenizing
Sessile Species & Rectangular Biome	Non-homogenizing	>	Homogenizing
Taxonomic Group & Spatial Scale	High z	>	Low z

Table 2 summarizes the major trends demonstrated by our calculations. These trends hold for all but one of the eight possible combinations of biome shape, species mobility, and climate warming scenario addressed by our model. The scenario that we have yet to analyze, due to mathematical complexity, is sessile species loss following a homogenizing temperature rise in a triangular biome.

The gross trends outlined in Table 2 are applicable to many pertinent issues involving global climate change and global biodiversity. Our analyses indicate that species loss following climate change is a function of biome shape. Species may be able to survive under crowded conditions for a limited period of time following climate warming; however, competition will eventually result in a new equilibrium species richness. Under equilibrium conditions, we predict that biomes that shrink with latitude towards the poles will suffer comparable or higher species losses than those with rectangular shapes.

Species loss following climate change is also a function of species type. We predict that sessile species will suffer greater losses than mobile species. In addition, for both mobile and sessile species, the magnitude of species loss depends on the SAR exponent z. As z increases, the fraction of species lost increases. The SAR exponent z is a function of two variables: spatial scale and taxonomic group (Rosenzweig 1995, Harte 1999b). Values of z, which are bounded between 0 and 1, often decrease with increasing spatial scale. Thus, our model predicts that smaller biomes will generally lose a greater fraction of their species than larger biomes. For a particular biome and spatial scale, z is a function of taxonomic group. Taxonomic groups with large values of z will suffer greater losses than those groups characterized by a smaller z.

Finally, the magnitude of species loss will depend on whether increases in surface temperature are uniform across the biome. Our model predicts that climate change magnified towards the poles leads to greater mobile species loss, yet less sessile species loss, in rectangular biomes. In triangular biomes, climate change magnified towards the poles leads to less mobile species loss than a uniform climate change. Hence our model predicts that relative effects of homogenizing versus non-homogenizing temperature change depend on both species type and biome shape.

4.2 Beyond Our Model

We have explored the influence of biome shape, biome size, species mobility, and geographic temperature warming variability on species loss by applying the SAR and EAR to several simplified scenarios. In calculating the fraction of species lost for each example, our intent was to observe trends in the influence of the aforementioned parameters, not to predict precise values. Several of the numerous complex interactions in the physical and biological world that are not adequately addressed in our analyses are summarized below. Brown and Lomolino (1998) and Harte et al. (1992) present a more detailed discussion of these issues.

We have assumed that species are either strictly sessile or mobile, but in reality, biomes are comprised of species with a wide range of dispersal and colonization ability. All but the most efficient dispersers will experience a time lag before extensive colonization is possible and hence, in the short term, experience range diminishment. Even highly mobile dispersers may be incapable of shifting their range if faced with physical barriers such as mountains, cities, roads, rivers, or farmland. In addition, the distribution of many mobile species will be limited by the distribution of relatively sessile species that provide their habitat.

Because species have different dispersal abilities and respond individually to various ecological forces, communities will tend to fragment. Differential shifting will change relative species abundances, create novel species associations, and potentially stress species by forcing them to cope with new predators, competitors, and diseases. Species dependent on mutualistic interactions may not be able to shift in synchrony with their mutualists. The combination of these effects will induce further indirect effects on biodiversity.

Community composition and the distribution of individual species are limited by environmental parameters and disturbance regimes that are highly influenced by geospheric processes and conditions. Alterations in the intensity and frequency of extreme events such as wildfires, flooding, droughts, blizzards, and hurricanes may produce more rapid changes in ecosystem community structure than the direct effect of increased temperature. Significant changes in the phenology of snowmelt as well as changes in soil structure and chemistry may also have substantial widespread ecological repercussions.

The biological consequences of climate warming cannot be viewed in isolation from other anthropogenic stresses such as deforestation, acid deposition, tropospheric pollution, and stratospheric ozone depletion. Plants and animals that are weakened by one stress may become more vulnerable to other stresses. Hence, the stress on biota from climate warming will likely be exacerbated by stresses from other anthropogenic sources.

The complex web linking climate change and biological systems makes it impossible to predict the magnitude of future impacts on biological species diversity with certainty. We have outlined a few of the interactions that will play a major role in the response of ecosystems that are not directly addressed in our analyses. In addition to the shortcomings already presented, our model simplifies many realistic features of the environment that will influence ecosystem dynamics such as topography and edge effects. Also, by calculating the fraction of species lost, we have neglected impacts to genetically unique populations, an equally important aspect of biological diversity.

5. SUMMARY

Climate warming will contribute to global biotic homogenization. We have explored how the shapes of biomes, the nature of global warming, and species characteristics such as mobility and taxonomic group impact the magnitude of species loss. Under equilibrium conditions, biomes that taper in area towards the poles will suffer comparable or greater species loss than those that do not. If climate change is more intense at the poles than at the equator (homogenizing), mobile species loss in rectangular biomes will be greater than if climate warming was uniform with latitude. If climate warming is uniform across a biome (non-homogenizing), sessile species loss in rectangular biomes and mobile species loss in triangular biomes will be greater than if climate change intensifies towards the poles. Species loss will be magnified for regions and taxonomic groups characterized by a large SAR exponent z. Finally, within a specified taxonomic group, sessile species will suffer more extinctions on average than mobile species. Because our methods and assumptions are based on a simplified model, it is reasonable to expect that losses in species richness may be exacerbated beyond the projections of our model by the effects of species interactions, geospheric-mediated processes, and anthropogenic stresses other than global warming.

ACKNOWLEDGEMENTS

Support from the Class of 1935 Endowed Chair and the Toxic Substances Research and Training Program at the University of California were instrumental in pursuing this work. We thank Abra Bentley, Lara Kueppers, Bill Riley and Scott Saleska for useful comments on the manuscript.

REFERENCES

Brown, J.H. and M.V. Lomolino, 1998. Biogeography. Sinauer Associates, Inc.

Harte, J., Torn, M., Jensen, M., 1992. The nature and consequences of indirect linkages between climate change and biological diversity In: Global Warming and Biological Diversity (Peters, R.L., Lovejoy, T.E., eds.). Yale University Press, London.

Harte, J. and A. Kinzig, 1997. On the implications of species-area relationships for endemism, spatial turnover, and food web patterns. Oikos 80(3):417-427.

Harte, J., Kinzig, A., Green, J.L., 1999a. Self-similarity in the distribution and abundance of species. Science 284:334-336.

Harte, J., McCarthy, S., Taylor, K., Kinzig, A., Fischer, M.L.. 1999b. Estimating species-area relationships from plot to landscape scale using species spatial-turnover data. Oikos 86(1): 45-53.

Harte, J., Kinzig, A., Green, J.L., 1999c. Response to: On the distribution and abundance of species (Maddux, R.D. and K. Athreya). Science 286:1647a.

Intergovernmental Panel on Climate Change, 1996. Climate Change 1995. Report of Working Group I, (Houghton, J.T. et al. eds.). Cambridge University Press, Cambridge.

Kinzig, A. and J. Harte, in press. Implications of endemics-area relationships for estimates of species extinction. Ecology.

Kunin, W.E., 1997. Sample shape, spatial scale and species counts: Implications for reserve Design, Biological Conservation 82:369-377.

MacArthur, R.H., 1972. Geographical Ecology. Harper and Row, New York.

McDonald, K.A. and J.H. Brown, 1992. Using montane mammals to model extinctions due to global change. Conservation Biology 6:409-415.

Parmesan, C., 1996. Climate and species' range. Nature 382:765-766.

Parmesan, C., Ryrholm, N., Stefanescu, C., Hill, J.K., Thomas, C.D., Descimon, H., Huntley, B., Kaila, L., Kullberg, J., Tammaru, T., Tennet, W.J., Thomas, J.A., Warren, M., 1999. Poleward shifts in geographical ranges of butterfly species associated with regional warming. Nature 399:579-583.

Peters, R.L., 1992. Conservation of biological diversity in the face of climate change. In: Global Warming and Biological Diversity (Peters, R.L. and Lovejoy, T.E., eds.) Yale University Press, London.

Peters, R.L. and J.D.S. Darling, 1995. The greenhouse effect and nature reserves: global warming would diminish biological diversity by causing extinctions among reserve species. Bioscience 35:707-717.

Rosensweig, M.L., 1995. Species Diversity in Space and Time. Cambridge University Press, Cambridge.

Thomas, C.D., Lennon, J.J., 1999. Birds extend their ranges northwards. Nature 399:213.

Chapter 10

The History and Ecological Basis of Extinction and Speciation in Birds

Peter M. Bennett[1], Ian P.F. Owens[2], Jonathan E.M. Baillie[1,3]
[1]*Institute of Zoology, Zoological Society of London, Regent's Park, London NW1 4RY, UK*
[2]*Department of Zoology and Entomology, University of Queensland, Brisbane, Queensland 4072, Australia*
[3]*Department of Biology, Imperial College, Silwood Park, Ascot SL5 7PY, UK*

1. INTRODUCTION

In 1995 John Lawton identified a fundamental unresolved problem in conservation biology. He wrote:

> "Some taxa (genera, tribes, families) are more extinction-prone than others, but we have no idea why......Although the reasons are often poorly understood, the implications are stark" (Lawton 1995, p. 159).

Birds provide a useful group in which to explore this question because they are exceptionally well studied in comparison to other animal groups. Comprehensive taxonomies and phylogenies have been developed that estimate the evolutionary relationships between species (Sibley and Ahlquist 1990; Sibley and Monroe 1990), and there is a wealth of information on the body size, ecology, life history and mating system of wild birds. Over 9,600 living species are recognized of which 11% are currently classified as being threatened with extinction by the IUCN (Collar et al. 1994). In the last decade considerable progress has been made in describing the geographical distribution of extinction risk, the first step in assigning priorities for the conservation of birds (Collar et al. 1994; Stattersfield et al.1998). Now we can be reasonably sure of which species are threatened with extinction and which regions have unusually high concentrations of threatened species. We also have an understanding of the relative importance of the main anthropogenic causes of threat to living birds - habitat loss, direct exploitation and introduced predators (Collar et al. 1994). However, while the importance of these extrinsic factors is well known, little progress has been made on the fundamental problem raised by John Lawton - why are some bird taxa vulnerable to extinction while others remain secure?

Recently, a number of workers have begun to apply robust comparative methods to the problem of understanding the evolutionary and ecological processes that underlie variation in extinction risk (Gaston and Blackburn 1995; Bennett and Owens 1997; Russell et al. 1998; Hughes 1999; Lockwood et al. 2000; Owens and Bennett submitted), species richness (Barraclough et al. 1995; Mitra et al. 1996; Møller and Cuervo 1998; Owens et al. 1999), rate of speciation (Cardillo 1999) and introduction success (Simberloff 1992; Cassey submitted; Lockwood et al. 2000) across bird taxa. By applying rigorous statistical methods that address the problems of phylogenetic non-independence, autocorrelation between variables, random variation and degree of explanatory power, a number of these studies are providing useful insights into those factors that are most likely to resolve Lawton's conundrum.

In this chapter we discuss four main questions. First, can the pattern of past bird extinctions help us to understand variation in extinction risk among living birds? Second, how do we identify taxa that are either unusually vulnerable to extinction or unexpectedly secure? Third, are there biological correlates of this variability that help us to understand why some bird groups are especially susceptible to extinction-risk? Fourth, are patterns of extinction and speciation in birds determined by the same factors? Throughout our emphasis is on answering the fundamental question: is each threatened bird species at risk because of a unique combination of factors, or are there general principles underlying the pattern of global extinction?

2. PREHISTORIC EXTINCTIONS

Can the patterns of prehistoric (last 10,000 years) and historic (last 400 years) bird extinctions help us to understand variability in extinction-risk among living birds? The aim of this section is to compare the pattern of variation revealed by the fossil and historical records of extinction with the pattern of variation shown by the IUCN list. This comparison is worth making because it has been suggested that both the rates and the causes of extinction may have changed over recent evolutionary time, although this is hard to demonstrate quantitatively.

Our knowledge of prehistoric bird extinctions is dominated by inferences drawn from the sub-fossil record. Over 80% of post-glacial bird extinctions (prior to those seen in the last 400 years) have been recorded on oceanic islands and they continue to be detected on most islands that are surveyed

(James 1995). The majority of the documented pre-historic oceanic island bird extinctions have occurred in the Pacific islands and the Caribbean, while comparatively few prehistoric extinctions are known from the Atlantic or Indian Oceans (Milberg and Tyrberg 1993; McCall 1997; Steadman 1995; Steadman et al. 1999). However, this pattern may simply reflect the large number of bird species in the Pacific, differences in sampling effort between oceans, or differences in the survival of evidence. The geology and climate of islands, for example, may influence the likelihood of discovering sub-fossil remains.

If taken at face value, this evidence suggests that the rate of avian extinction over the past 10,000 years - the time period in which most islands were colonized by humans - has been much higher than standard estimates of the 'background rate of extinction' in birds (Pimm et al. 1995). It has been estimated that the extinction of one species per year is the background rate for all taxonomic groups (May et al. 1995), although it may be slightly higher for vertebrates (Ehrlich et al. 1977; Martin 1993). Even if avian extinctions were 10 times the estimated average rate for all species, we would expect less than one extinction every hundred years for all birds. Although the true magnitude of historic and prehistoric avian extinctions is unknown (Bibby 1995; Pimm et al. 1995), one well quoted extrapolation from excavations of tropical Pacific islands indicate that more than 2,000 species of bird have gone extinct during the past 10,000 years (Steadman 1991, 1995). This is equivalent to a rate of extinction of one species every five years, which is considerably higher than the standard 'background' estimate of one avian extinction every 100 to 1,000 years. Indeed, if the sub-fossil records of the Pacific islands are representative of the overall pattern of global avian extinction, it suggests that approximately one fifth of all bird species have gone extinct over the past 10,000 years (Milberg and Tyrberg 1993). Moreover, it is becoming increasingly fashionable to lay the responsibility of these extinctions at the door of humans.

When we review the proposed causes of known Holocene bird extinctions we find that humans have been implicated in many of them (Olson 1977; Diamond 1982, 1984, 1989; Olson and James 1982; Cassels 1984; Steadman and Martin 1984; Dye and Steadman 1990; Milberg and Tyberg 1993; Pimm et al. 1995; Benton and Spencer 1995; Wragg 1995; James 1995; Steadman 1995; Steadman et al. 1999). While it is difficult to determine the major causes of prehistoric avian extinction, it is widely believed that exploitation by humans, introduced species and to a lesser extent habitat destruction have all been major contributing factors. However, this view is largely based on conjecture and opinion rather than hard evidence. Most known Holocene bird extinctions are believed to have been flightless species found on islands (see estimates in Steadman 1995).

Flightless birds as well as colonial sea birds would appear to have been obvious targets for early colonists and relatively easy to capture for food. Naturally invasive or anthropogenically introduced species may lead to increased predation, competition, alteration of habitat and disease (Diamond 1984). Although the number of species known to have been introduced by prehistoric humans to islands is relatively small, it has been argued that the impact of the human commensal travelers such as the Polynesian rat *Rattus exulans* (Roberts 1991), dogs, pigs and poultry (Keegan and Diamond 1987), and lizards and snails (Kirch 1982), was significant. The role of habitat loss in prehistoric extinctions may have varied greatly. While it is difficult to assess the magnitude of habitat alterations as a result of early colonists, prehistoric human modification of the landscape is evident on islands such as Puerto Rico, Madagascar (Steadman 1995) Easter Island (Fenley and King 1984), New Zealand (Holdaway 1989) and Hawaii (Olson and James 1982).

Although it is near-impossible to unambiguously diagnose the cause of extinction from sub-fossil evidence alone, the literature suggests that the major causes of extinction during the Holocene were, in order of importance: introduced predators and diseases; direct persecution by humans; and habitat-loss. This is the opposite order of importance to that seen among contemporary threatened species.

3. HISTORIC EXTINCTIONS AND CONTEMPORARY THREATENED SPECIES

Although many historic extinctions must have passed unnoticed (Olson 1977), bird extinctions over the past 400 years are the best documented of any class (Jenkins 1992). 110 bird extinctions have been documented over this period (Collar et al. 1994). This suggests a rate of extinction of approximately one species every four or five years - very similar to the estimate based on the subfossil record, but much higher than the standard background rate of one extinction every 100 or 1,000 years. Once again, however, the database on historical extinctions is dominated by island-dwelling species - with over 88% of historical extinctions being from oceanic or continental islands and only 12% on the mainland (Collar et al. 1994; Stattersfield et al.1998).

Historic extinctions and currently threatened species show a similar geographical pattern to prehistoric extinctions with the majority found in the Pacific. However, the second most extinction prone region is the Indian Ocean, followed by the Caribbean. It is not surprising that the majority of threatened and recently extinct birds are found in the Pacific, as this region harbors most of the oceanic islands of the world. Nevertheless, it is interesting that so many known recently extinct and currently threatened species have occurred on islands of the Indian Ocean. This may be a result of the so-called 'extinction filter effect' (Pimm et al. 1995; Balmford 1996). Many of the islands in this region have only recently been colonized (over the last 2,000 years) by humans. It is possible that species that are particularly susceptible to human-related threats will have already been 'filtered' from the islands that have been settled for long periods of time, whereas recently colonized islands are currently experiencing this extinction filtration effect. This relationship may be observed in the Pacific islands where the regions that have most recently been colonized have a higher proportion of threatened and recently extinct species (Pimm et al. 1995). It is also supported by the fact that the islands with the greatest number of known recent bird extinctions (New Zealand, Hawaii, Mauritius, Rodrigues, and Reunion) have all been colonized within the past 1,500 years.

While oceanic extinctions dominate the historic record, the majority (54%) of currently threatened species are found on the mainland (Collar et al. 1994). However, a much greater proportion of island birds are threatened than mainland birds (Johnson and Stattersfield 1990); 28% of island birds versus 8% of mainland birds are threatened. This shift of extinction risk from islands to mainland (King 1985, Mountfort 1988, Johnson and Stattersfield 1990) does not mean that island species are any less threatened, only that proportionally more mainland species are now at risk of extinction. A recent study by Manne et al. (1999) found that among passerine birds, mainland species are more threatened than island species when they corrected for range size (island species have inevitably smaller ranges than mainland ones).

Five families account for about half the documented prehistoric bird extinctions on islands (Milberg and Tyrberg 1993; McCall 1997; Steadman 1995; Steadman et al. 1999) and roughly half the known historic documented extinctions. Four of these families are the same for prehistoric and historic extinctions. These are the rails (Rallidae), Hawaiian honeycreepers (Drepanididae), ducks (Anatidae), and pigeons (Columbidae). However, it should be noted that these are relatively large families and that there has apparently been a disproportionately high number of extinctions among species poor taxa in the historic record. For both known historic extinctions (Russell et al. 1998) and currently threatened species (Bennett and Owens 1997), the distribution of extinct or threatened status has been shown to be non-random across families (see below).

We have collated information on the most likely causes of extinction for 79 of the 110 historically documented extinctions (data from Greenway 1967; Prestwich 1976; Bengston 1984; Collar and Stuart 1985; Mountfort

1988; Brouwer 1989; Johnson and Stattersfield 1990; Clements 1991; Gill and Martinson 1991; Jenkins 1992; Collar et al. 1994.) We have compared these findings with the main causes of threat to living birds (see Figure 1). The data for living birds is for the 1,111 species listed as Critical, Endangered or Vulnerable to extinction by Collar et al. (1994). If our estimates of the number and causes of recent extinctions are realistic (see below for reasons why this is likely to be an unsafe assumption) then it appears that in the recent past, introduced species were the dominant cause of extinction, but now habitat loss is the greatest threat (see also Johnson and Stattersfield 1990). If this shift in the dominant process of extinction over time is real it may help to explain why the extinction risk of birds is now apparently increasing on the mainland. Manne et al. (1999) have also argued that island birds have been unusually vulnerable to the human introduction of previously absent predators and disease, but mainland forms while less ecologically naïve are also highly threatened by habitat loss.

Once more, this analysis suggests that the order of extinction threats over the past 400 years is different from that observed today. The most common sources of extinction threat for the historical extinctions were, in order of importance, introduced predators and diseases; direct human exploitation; and habitat-loss. This is the same order suggested by the extrapolations from the sub-fossil records of Pacific islands, but the opposite to that seen among contemporary threatened species.

4. PROBLEMS WITH COMPARING PAST AND PRESENT EXTINCTION PATTERNS

On first inspection, therefore, the sub-fossil and historical evidence suggests that there may have been an increase in the rate of avian extinction over the last 10,000 years or so and, although the major causes of extinction during that period were anthropogenic, they were a subtly different set of anthropogenic causes to those experienced by contemporary threatened species. While habitat-loss is now the most common cause of extinction among birds, it seems that direct human persecution and introduced predators have been much more important in recent historical times.

But all of these conclusions must be treated with the utmost caution. As we have already hinted, the prehistoric, historic and contemporary estimates of extinction-risk are derived from very different sources. We know of at least five sources of systematic variation in the way that extinction estimates are calculated across these three periods and we will discuss them briefly.

10. Extinction and Speciation in Birds

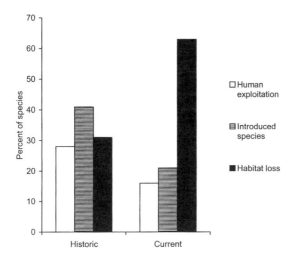

Figure 1. Frequency histogram comparing presumed main causes of extinction in historic times (n = 79 species in last 400 years) with current threats to living birds (n = 1111 species).

1. Species concepts vary across periods. In the study of sub-fossils, 'morpho-species' must be identified on the basis of fragments of bone, feathers and skins. Depending on the practitioners involved, such a method may lead to either more or less species than identified by colonial collectors who had access to intact specimens (historical approach) or contemporary taxonomists who use an arsenal of morphological, ecological, behavioural, geographical, molecular and statistical approaches to identify species.
2. Extinction concepts vary across information sources. Extinction is diagnosed from the sub-fossil record by the absence of bones. However, a lack of bones in a particular deposit could also be due to changes in suitability of conditions for sub-fossilisation or changes in the distribution of the organism in question. Extinction is diagnosed from historical records largely on the basis of colonial explorers and collectors. Extinction among contemporary species is estimated as a probability function based on the population size, stability and range of the species in question. In cases of likely extinction, one or more teams may endeavour to conserve the species in question. It would be hard to make an argument that all these approaches were equal.
3. The method of inferring the cause of extinction varies across periods. The causes of prehistoric extinctions must be inferred from indirect evidence associated with broad patterns of human colonisation, sub-fossil remains associated with ancient human habitation, and patterns of vegetation

changes inferred from pollen grain analysis. The cause of historical extinctions must be inferred from the records maintained by collectors and explorers and anecdotal evidence passed down by word-of-mouth. The causes of contemporary extinction-risk are often based on detailed, sometimes experimental, study of the ecology and population dynamics of individuals populations over time. Once again, it would be hard to make an argument that all these approaches were equal.

4. The survival of evidence varies across periods. For a prehistoric extinction to occur in our databases a delicate pattern of evidence must be both preserved and recognised. For historical extinctions to occur in our databases they must occur in a region frequented by Western cultures and be of a species that was already known to science. For example, large body size may influence the likelihood of discovery. Contemporary extinctions, on the other hand, are far more likely to make it to the databases, given that there is no need to rely on the uncertain processes of sub-fossilisation, rediscovery or haphazard reporting. Indeed, there are international organisations established to record the probability of extinction among living birds.

5. The biogeographic region of study varies across information sources. Almost all evidence of prehistorical and historical patterns of extinction is based on island-dwelling birds. There are a host of theoretically plausible reasons to suspect that island-dwelling birds are more susceptible to extinction than their mainland-dwelling relatives. Also, given the unusual tameness and flightlessness among island races, they should be more vulnerable to novel ground-dwelling predators. Together, these factors suggest strongly that analyses based on island-dwelling species alone will overestimate the rate of extinction and mis-assign the importance of different causes of extinction. Unlike the problems outlined above which relate to data inconsistency, this focus on island-dwelling birds is of particular interest because it is biological variation that requires explanation.

Given this catalogue of systematic differences in the manner in which extinction rates are calculated, we must remain cautious in drawing conclusions about changes in the rate or the pattern of extinction over time. This problem of systematic biases in both the survival of evidence and the geographic distribution of sampling effort means that our inferences of the rates and causes of past extinctions are much weaker than our knowledge of current threats to birds. Some authors have excluded recent extinctions from their studies for these reasons (e.g. Bennett and Owens 1997) while others have included them (e.g. Bibby 1995; Russell et al. 1998). The important point is to recognize that these systematic biases exist.

10. Extinction and Speciation in Birds

So, what is the answer to our question posed in the Introduction - can the pattern of past bird extinctions help us to understand variation in extinction risk among living birds? If we take the pre-historic and historic records at face value, it seems that the current pattern of extinction risk among birds is strikingly different from that which has occurred in the recent past. The extinction rate is now much higher, and the causes of extinction-risk are different - habitat loss is more important, while direct persecution is less so. In other words, both the magnitude and the mechanism of avian extinction appear to have changed over the last 10,000 years. Again, if this is taken at face value it suggests that (1) an understanding of prehistoric and historic extinction in birds will be of only limited value in understanding the current pattern of risk, (2) some regions and some lineages may already have gone through an extinction filter, (3) extinction via habitat loss is the overwhelming challenge to the survival of both mainland and island birds (although we must remain cautious because both the accuracy of the data and rigor of the analyses upon which this conclusion is based are questionable). For both mainland and island taxa it appears that the challenge of habitat loss should be worked on with the greatest urgency by conservation agencies because many avian lineages, particularly mainland forms, are being exposed to this source of rapid anthropogenic change for the first time. It remains to be seen what the consequence of this new extinction filter (habitat loss) will have on avian diversity.

Of course, we have also argued that there are problems in taking the past extinction record at face value. An equally valid conclusion is to assume that the apparent temporal differences in extinction patterns in birds are due to sampling bias. If this were the case then an improved and less biased understanding of historical extinctions would be very valuable. So, we need to know the extent to which the differences in extinction pattern (for both rate and source of extinction) between historical and current times are due to bias and the extent to which they are due to biology. Further palaeontological and theoretical research will help to resolve this question, in particular, quantitative analyses such as simulation techniques comparing mainland and island extinction patterns would be valuable.

5. VARIATION IN EXTINCTION RISK AMONG LIVING SPECIES

Living bird species and families are not equal in their risk of extinction. Some birds such as the Californian condor Gymnogyps *californianus*, bald ibis *Geronticus* eremita, and night parrot *Geopsittacus* occidentalis teeter on the brink of oblivion, while many other species appear secure. Likewise, some families such as the parrots (Psittacidae) and pheasants (Phasianidae) appear to contain an unusually high proportion of threatened species while others such as the woodpeckers (Picidae) and cuckoos (Cuculidae) have low numbers of threatened species (Bennett and Owens 1997). Why is this so? What makes some groups of species more prone to extinction than others?

Some authors have claimed that there are a number of obvious characteristics shared by bird species that have become extinct. For example, most known prehistoric and historic bird extinctions have occurred on islands (see above) where species have evolved in the absence of many competitors and predators. The majority of known Holocene bird extinctions were believed to have been flightless (see Steadman 1995) and roughly 30% of birds that have recently become extinct were flightless or near flightless (see references above). If follows that many of these species would have nested on the ground so that eggs and adults would have little defense against introduced predators and persecution by man. However, such an analysis is probably biased by incomplete data on bird extinctions and uneven sampling effort (see above). In addition, the vast majority of currently threatened bird species can fly.

A large number of intrinsic characteristics have been hypothesized to be associated with an increased risk of extinction in birds and many of them are strongly inter-correlated. These include: large body size (Pimm et al. 1988; Gaston and Blackburn 1995; Bennett and Owens 1997), larger members of guilds (Terborgh 1974), specialists (Bibby 1995), reduced fecundity (Pimm et al. 1988; Bennett and Owens 1997), high trophic levels (Terborgh 1974; Diamond 1984), colonial nesters (Terborgh 1974), migratory species (Pimm et al. 1988), heightened secondary sexual characteristics (McLain et al. 1995; Sorci et al. 1998), low genetic variability (Frankham 1997; 1998), species poor lineages (Russell et al. 1998), increased evolutionary age (Gaston and Blackburn 1997), small population size (MacArthur and Wilson 1967), and species with greater population fluctuation (Leigh 1981; Pimm et al. 1988). The majority of these hypotheses remain to be rigorously tested.

5.1 Null models and the taxonomic distribution of extinction-risk

It is important to recognize that explaining variation in extinction risk among living birds is a two step process (Bennett and Owens 1997). First, we must establish whether variation in the threat of extinction is randomly distributed among avian families. We need to know the taxonomic

10. Extinction and Speciation in Birds

distribution of extinction risk because it is possible that variation in susceptibility to extinction is solely due to external factors such as human disturbance or catastrophic events. If this were true, variation in extinction-risk may be randomly distributed among species (see Raup 1991) - any species that is affected by external factors will be threatened by extinction, irrespective of its biology. Of course, the most likely scenario is that variation in extinction-risk is due to an interaction between intrinsic and extrinsic factors. Second, if extinction risk shows a non-random distribution then we must test which candidate biological factors are correlated with variation in extinction risk.

We considered the first problem (Bennett and Owens, 1997) by developing a null model (MacArthur 1972) to test whether the distribution of threatened species among families could be explained by random allocation. To know what a random distribution of extinction-risk would look like we performed a simulation. Since 1,111 of all bird species are classified as threatened (categories Vulnerable, Endangered and Critical in Collar et al. (1994), we picked 1,111 species at random from the complete list of 9,672 bird species, noted which families they were from (using the classification of Sibley and Monroe 1990), and calculated the proportion of species in each family that had been randomly picked in this way. We then repeated this simulation many times and compared the predicted and observed distributions. We found that the two frequency distributions were significantly different. There are significantly more families that are more threatened than would be predicted by chance, and significantly more families that are less threatened than would be predicted by chance. Can we identify these families?

Bennett and Owens (1997) used the binomial distribution to identify those families which contained either a larger, or smaller, proportion of threatened species than would be expected by chance. Those families containing an unusually large number of threatened species are shown in Table 1. All families whose allocation of threatened species is either twice, or more, as high than expected (i.e. proportion threatened of 0.22 or more) are listed. The families which contain significantly more threatened species than expected using the binomial method are the parrots (Psittacidae), pheasants and allies (Phasianidae), petrels and allies (Procellariidae), rails (Rallidae) and pigeons (Columbidae). The only family which contains significantly fewer species than expected is the woodpeckers (Picidae). Other methods have also been applied to identifying taxonomic 'selectivity' with respect to extinction risk in birds and there is good agreement between them (Russell et al. 1998; Lockwood et al. 2000).

5.2 Ecological correlates of variation in extinction risk

Now that we have identified those lineages which are unusually vulnerable and those that are unexpectedly safe, we can test hypotheses that aim to explain variation among taxa in extinction-risk by its association with variation in biology. A large number of variables have been suggested to explain this non-random distribution of extinction-risk (see above). For each of these factors a plausible theoretical link to extinction can be established. However, convincing statistical associations between overall variation in extinction-risk and overall variation in these sorts of factor have proved depressingly illusive (Lawton 1995). Typically, variation in factors such as body size and fecundity explain less than 10% of the variation in extinction-risk, irrespective of the type of comparative technique employed (Gaston and Blackburn 1995; Bennett and Owens 1997). The only exceptions to this rule are analyses based on local population extinctions rather than global extinctions (Pimm et al. 1988), and analyses based on factors that autocorrelate with extinction-risk because they are used to calculate the index of extinction-risk in the first place, such as population size or range size. It has proved impossible, therefore, to assess quantitatively the relative importance of different factors in predisposing a lineage to high or low rates of extinction. We are left with the question, is each over-threatened lineage at risk because of a unique combination of factors, or are there general principles underlying the pattern of global extinction?

To address this question we collated data on 8,657 species from 95 avian families on extinction-risk and three well-known candidate ecological factors: body size, generation time and degree of habitat specialization (Owens and Bennett, submitted). We chose these variables for a number of reasons. First, there are well-established theoretical reasons to expect a correlation with extinction-risk. Second, data is widely available on these three variables. Finally, while it has been suggested that variation in these factors may be associated with variation in extinction-risk (e.g. Pimm et al. 1988), these factors are independent of the IUCN criteria used to classify threatened species. This is not the case with other characters such as measures of abundance or range size that are used to calculate the IUCN index of extinction-risk (Collar et al. 1994). Any correlation between variation in these sorts of factors and variation in extinction-risk, where extinction-risk is defined using the overall IUCN index, would be confounded by autocorrelation.

Extinction-risk was measured as the proportion of species in a family that are classified as being threatened by extinction (Collar et al 1994.). Unlike previous studies we analyzed each 'source' of extinction-risk (threat) separately and

10. Extinction and Speciation in Birds

Table 1. Highly threatened avian families (from Bennett and Owens 1997).

Family name	Common name	Richness	Number threatened	Proportion threatened
Apterygidae	Kiwis	3	3	1.00
Mesitornithidae	Mesites	3	3	1.00
Rhynochetidae	Kagu	1	1	1.00
Brachypteracidae	Ground-rollers	5	4	0.80
Menuridae	Lyrebirds	4	2	0.50
Orthonychidae	Logrunners	2	1	0.50
Picathartidae	Rockfowl	4	2	0.50
Casuariidae	Cassowaries	4	2	0.50
Gruidae	Cranes	15	7	0.47
Megapodidae	Megapodes	19	8	0.42
Phoenicopteridae	Flamingoes	5	2	0.40
Fregatidae	Frigatebirds	5	2	0.40
Odontophoridae	New World Quail	6	2	0.33
Callaeatidae	NZ Wattlebirds	3	1	0.33
Turnicidae	Buttonquails	17	5	0.29
Tytonidae	Tyto Owls	17	5	0.29
Spheniscidae	Penguins	17	5	0.29
Cracidae	Cracids	49	14	0.29
Procellariidae	Albatrosses	115	32	0.28
Phasianidae	Pheasants	177	45	0.25
Philepittidae	Asities	4	1	0.25
Heliornithidae	Sungrebes	4	1	0.25
Psittacidae	Parrots	357	89	0.25
Sittidae	Nuthatches	25	6	0.24
Ciconiidae	Storks	26	6	0.23
Pittidae	Pittas	31	7	0.23
Rallidae	Rails	142	32	0.23
Pelecanidae	Pelicans	9	2	0.22
Zosteropidae	White-eyes	96	21	0.2
Columbidae	Pigeons	309	55	0.18

[A] 9,672 species, data from Sibley & Monroe (1990)
[B] 1,111 threatened species (Vulnerable, Endangered and Critical), data from Collar et al. (1994)

found consistent correlations between variation in extinction-risk and variation in the candidate ecological factors (Figure 3). Extinction-risk incurred through habitat-loss, for example, was consistently associated with small body size, short generation time and habitat specialization (Figure 3b). Extinction risk incurred through persecution and introduced predators, on the other hand, was associated with large body size and long generation time, but was not associated with degree of habitat specialization (Figure 3c). Furthermore, these results remained qualitatively unchanged when we controlled for differing degrees of shared ancestry among the families by repeating our analyses on evolutionarily independent contrasts (Owens and Bennett, submitted).

These results suggest two general explanations for the paradox that variation in overall extinction-risk is only weakly correlated (see Figure 3a) with variation in theoretically plausible ecological factors (Lawton 1995). The first explanation is straightforward: some ecological factors are only associated with particular sources of extinction-threat. In the case of our analyses, the extent of habitat specialization is only associated with extinction-risk incurred via habitat-loss. The second explanation is subtler: some ecological factors are positively associated with one type of extinction-risk but negatively associated with another type of extinction-risk. In our analyses body size and generation time are examples of this sort of factor, both being positively associated with extinction-risk incurred via persecution or introduced predators but negatively associated with extinction-risk via habitat loss.

Our analyses identify a method for exploring the processes that underlie current extinction. By considering different sources of extinction-risk separately, we can test the relative importance of different ecological factors in predisposing species to extinction. The results obtained here confirm the prediction (Pimm et al 1988; Pimm 1991) that there are multiple routes to extinction among birds. One route, which has been described previously (Pimm et al 1988; Pimm 1991; Bennett and Owens 1997; Gaston and Blackburn 1995) is for relatively large bodied, slow breeding species to become threatened when an external factor, such as introduced predators or human persecution, disrupt the delicate fecundity-mortality balance. In our database, this route applies to groups such as the kiwis (Apterigidae), cassowaries (Casuariidae), megapodes (Megapodiidae), penguins (Spheniscidae), and petrels and allies (Procellariidae). A second route, which is poorly explored, is for small-bodied, fast-breeding, ecologically specialized species to become threatened by habitat loss. Such groups include the buttonquail (Turnicidae), hummingbirds (Trochilidae), trogons (Trogonidae), ovenbirds (Furnariidae), scrub-birds (Menuridae), thornbills and scrub-wrens (Pardalotidae), logrunners (Orthonychidae), shrikes (Laniidae) and New World wood-warblers (Fringillidae). Unenviably, a small number of families are prone to both sources of extinction-risk. These include the parrots (Psittacidae), rails (Rallidae), pheasants and allies (Phasianidae), pigeons (Columbidae), cranes (Gruidae) and white-eyes (Zosteropidae). It is no surprise that it is this last group of families that have previously been identified as being significantly over-prone to extinction (see Table 1, Bennett and Owens 1997; Russell et al. 1998).

6. PATTERNS AND CORRELATES OF SPECIES-RICHNESS

Similar techniques to those described above for extinction risk can also be used to understand patterns of species-richness in birds (Owens et al. 1999). The process of speciation is the only natural source of new species and the imbalance between current extinction and speciation rates is of great interest. Is it possible, for example, that some highly extinction prone lineages can replace species at a faster rate than more secure lineages?

We found strong evidence that the observed variation among bird families in species-richness is not simply a consequence of random branching patterns (Owens et al. 1999). According to Sibley and Monroe (1990) there are 9672 extant species of birds distributed among 145 taxonomic families. On average, therefore, each family contains about 67 species. The observed pattern is, however, far from even. Over half of the species are contained within just 12 species-rich families, each of which contains over 250 species. At the other end of the scale, almost half the families contain less than 10 species each, and account for less than 250 species between them. Using a similar null model approach to that described above for extinction risk, we found that there are far too many species-poor and too many species-rich families than would be expected from chance mechanisms alone (Owens et al. 1999). These results support the idea that it is worth seeking correlates of species-richness among birds. However, our results did not support the traditional explanation that small body size and fast life history are correlated with high species richness. Instead, we found that the main correlates were indices of the strength of sexual selection and ecological and geographic variables. For example, species-richness is associated with indices of ecological generalism and dispersal ability. These results support Rosenzweig's (1995) 'geographic' model of diversification whereby the chances of a lineage becoming species-rich is closely associated with its chances of finding, and then successfully colonizing, new areas. Dispersive forms that can cope with a variety of conditions will successfully colonize new areas, will have a large geographic range and are, therefore, likely to become subdivided by geographic isolating mechanisms.

Analyses based on methods designed to increase the number of independent contrasts in species-richness analyses largely confirmed these results (Owens and Bennett, unpublished data). In addition, however, they revealed some differences in the correlates of species-richness between clades. The correlation between species richness and sexual dichromatism, for example, is particularly strong among passerine families but particularly

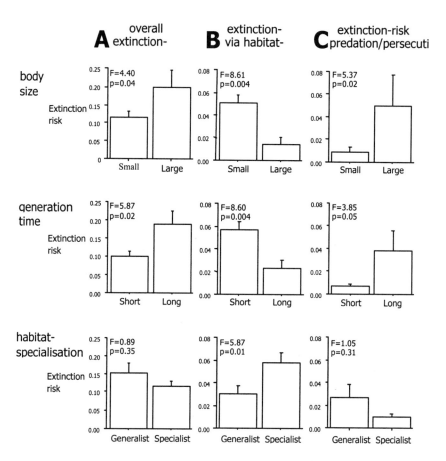

Figure 2. Associations between variation in ecology and variation in sources of extinction-risk across bird families (from Owens and Bennett, submitted). Extinction-risk (the Y-axis in each case) is the proportion of species in a family that are classified as being threatened by extinction. (A) Overall extinction-risk versus body weight, generation time, and degree of habitat specialisation, respectively. (B) Extinction-risk incurred via habitat-loss alone versus body weight, generation time, and degree of habitat specialisation, respectively. (C) Extinction-risk incurred via introduced predators or direct human persecution alone versus body weight, generation time, and degree of habitat specialisation, respectively. Small and Large body size refer to below and above 1,000g, respectively. Short and long generation time refer to below and above 18 months at first breeding, respectively. Generalist and Specialist refer to more than and less than one habitat-type category used for breeding, respectively. All analyses are based on family-typical values for 95 avian families. Error bars show standard errors, statistics show results of one-way ANOVAs. Degrees-of-freedom in all ANOVAs =1, 93.

weak among the Ciconiiformes. This suggests that sexual selection may be the critical factor leading to the huge species-richness of the passerines. Among the Ciconiiform families, on the other hand, the strongest correlates of species-richness are geographic range size and extent of geographic fragmentation. This ties in with the observation that the Ciconiiformes, which are largely waterbirds, radiated at approximately the same time as Gondwanaland was breaking up (Cotgreave and Harvey 1994) - and many new islands were being created.

The traditional view is that extinction and speciation are simply opposite sides of the same biological mechanism. That is, they are both determined by the same factors, of which body size and life history are of key importance. Large body size and slow life histories lead to high extinction and low speciation. However, we have demonstrated that, although body size and life history are important in determining extinction risk among birds (see also Gaston and Blackburn 1996; Bennett and Owens 1997), they are relatively unimportant in terms of determining species-richness. Species-richness, on the other hand, appears to be largely determined by sexual selection and geographic factors such as range size (Owens et al. 1999), which may not be correlated with extinction-risk among living bird families in a straightforward manner (Owens and Bennett unpublished data). Thus extinction and speciation may not be opposite sides of the same biological mechanism.

Given these contrasting patterns of covariance for extinction-risk and species-richness, we suspect that lineages could combine high-extinction with high-speciation, with the high extinction-risk being naturally offset by the high rate of cladogenesis. At any one moment in time, such lineages would appear very species rich but a large proportion of species would be vulnerable to extinction. Possible candidates among bird families include the parrots (Psittacidae), rails (Rallidae), pigeons (Columbidae) and pheasants (Phasianidae). All of these lineages contain more than twice as many threatened species as expected by chance (Bennett and Owens 1997), but given that they are so species rich perhaps this is the natural situation? This possibility requires further study (see Owens et al. 1999), however, it is likely that anthropogenic change is leading to an imbalance in the processes of extinction and speciation in these families. It is interesting to note that some of these naturally extinction-prone families show the phylogenetic pattern suggested by Nee et al. (1996) of a large increase in the net rate of cladogenesis over time. This suggests that they are either subject to a high rate of speciation and a high rate of extinction, or that the balance between speciation and radiation has changed dramatically over time for these groups but not for others.

7. CONCLUSIONS

We have identified bird families that contain an unusually high number of threatened species while some are unexpectedly safe. We have also found biological correlates that help to explain this pattern. While the highly threatened families in Table 1 are of great concern, perhaps we should not view them all in the same way. Families like the kiwis, mesites, kagu, logrunners and so on, represent lineages in which the current high level of threat is probably something very new in evolutionary terms. They are naturally species-poor and cannot sustain the current level of threat. Members of other families, however, such as the parrots, rails and pigeons, have perhaps always been on the threatened species lists. Fortunately for them, they are also good at speciation, and a large number of 'threatened' species characterized by small, restricted range populations is to be expected. In addition, these families are speciose with large mainland source populations, are good at colonizing geographically restricted ranges, and are the most common endemics on islands. However, in the recent past these families have been exposed to rapid anthropogenic change and this combined with their naturally high extinction rates means that we cannot be complacent with respect to their conservation - it is extremely unlikely that current rates of extinction and speciation balance one another.

We are less confident of being able to compare quantitatively patterns of extinction over time. We should exercise caution when making comparisons between periods because it is likely that there are systematic differences in the way that extinction rates and extinction patterns are estimated from sub-fossil, historical and contemporary evidence. The estimates of extinction rate based on sub-fossil and historical evidence are at least an order of magnitude higher than those based on traditional paleontological approaches. Is this because the rate is now ten times higher than it was, or is it because of differences in extrapolation? Similarly, comparisons between contemporary and either prehistoric or historic evidence suggests that the relative impact of direct human persecution and introduction of predators on persistence has changed over the previous 10,000 years. Is this because (1) habitat-loss has now taken over as the major source of extinction risk, (2) we have already killed all the vulnerable species (the 'extinction-filter hypothesis'), or (3) the prehistoric and historic data are systematically biased because they are based almost entirely on island-dwelling species? Or is it a combination of these and other factors. These questions require rigorous critical investigation.

ACKNOWLEDGEMENTS

We thank Susan Scott for help with data collection, P. Agapow, K. Arnold, T. Barraclough, S. Blomberg, L. Bromham, M. Cardillo, M. Charleston, S. Clegg, G. Cowlishaw, M. Cunningham, P. Dwyer, A. Goldizen, T. Hamley, B. Holt, N. Issac, J. Kikkawa, P. Harvey, W. Jetz, R. Lande, G. Mace, M. Minnegel, C. Moritz, S. Nee, V. Olson, S. Pimm, A. Purvis, A. Rambaut, C. Schäuble, A. Smyth, A. Stattersfield, N. Stork and A. Thomas who have discussed ideas, commented on papers, and/or provided unpublished data for our studies on extinction and species-richness in birds. We thank NERC, the Australian Research Council, the University of Queensland, the Overseas Research Studentship, and the Harshman Fellowship for financial support.

REFERENCES

Balmford, A. 1996. Extinction filters and current resilience: the significance of past selection pressures for conservation biology. Trends in Ecology & Evolution 11:193-196.

Barraclough, T.G., P.H. Harvey, & S. Nee. 1995. Sexual selection and taxonomic diversity in passerine birds. Proc. Roy. Soc. Lond. B 259:211-215.

Bengtson, S. 1984. Breeding ecology and extinction of the Great Auk (Pinguinus impennis): anecdotal evidence and conjectures. Auk 101.

Bennett, P. M. and I. P. F. Owens. 1997. Variation in extinction risk among birds: chance or evolutionary predisposition? Proceedings Of the Royal Society Of London Series B-Biological Sciences 264:401-408.

Benton, T. G., and T. Spencer. 1995. Henderson Island prehistory: colonization and extinction on a remote Polynesian Island. Biological Journal of the Linnean Society 56:377-404.

Bibby, C. J. 1995. Recent past and future extinctions in birds. In: Extinction Rates (J. H. Lawton and R. M. May, eds.), Oxford University Press, Oxford, Pp: 98-110.

Brouwer, J. 1989. An annotated list of the rare, endangered and extinct birds of Australia and its Territories. Conservation Committee Royal Australian Ornithologists Union, Victoria.

Cardillo, M. 1999. Latitude and rates of diversification in birds and butterflies. Proceedings Of the Royal Society Of London Series B-Biological Sciences 266:1221-1225.

Cassels, R. 1984. The role of prehistoric man in the faunal extinctions of New Zealand and other Pacific islands. In: Quaternary extinctions: a prehistoric revolution (P. S. Martin and R. G. Klein, eds.), University of Arizona Press, Tucson.

Cassey, P. (submitted). Variation in the success of introduced avifauna. 28pp.

Clements, J. F. 1991. Birds of the world. A check list. Ibis Publishing Company, California.

Collar, N. J., M. J. Crosby and A. J. Stattersfield. 1994. Birds to Watch 2 - The World List of Threatened Birds. BirdLife International, Cambridge, UK.

Collar, N. J. and S. N. Stuart. 1985. Threatened birds of Africa and related islands. The IUCN/ICBP red data book. IUCN/ICBP, Cambridge, UK.

Cotgreave, P. and P.H. Harvey. 1994. Phylogeny and the relationship between body-size and abundance in bird communities. Functional Ecology 8:219-228.

Diamond, J. A. 1982. Man the exterminator. Nature 298:787-789.

Diamond, J. M. 1984. "Normal" extinctions of isolated populations. In: Extinctions (M. H. Nitecki, ed.), University of Chicago Press, Chicago, Pp: 191-246.

Diamond, J. M. 1989. The present, past and future of human-caused extinctions. Philosophical Transactions of the Royal Society, London, Series B 325:469-477.

Dye, T. and D. W. Steadman. 1990. Polynesian ancestors and their animal world. American Scientist 78:207-215.

Ehrlich, P. R., E. A. H., and J. P. Holdren. 1977. Ecoscience: population, resources, environment. W. H. Freeman, San Francisco.

Fenley, J. R. and S. M. King. 1984. Late Quaternary pollen records from Easter Island. Nature 307:47-50.

Frankham, R. 1997. Do island populations have less genetic variation than mainland populations? . Heredity 78: 311-327.

Frankham, R. 1998. Inbreeding and extinction: island populations. Conservation Biology 12:665-675.

Gaston, K. J. and T. M. Blackburn. 1995. Birds, body size and the threat of extinction. Philosophical Transactions of the Royal Society of London Series B- Biological Sciences 347:205-212.

Gaston, K. J. and T. M. Blackburn. 1997. Evolutionary age and extinction risk in global avifauna. Evolutionary Ecology 11:557-565.

Gill, B., and P. Martinson. 1991. New Zealand extinct birds. Random Century New Zealand Ltd., Hong Kong.

Greenway, J. C. 1967. Extinct and vanishing birds of the world. Dover Publications, New York.

Holdaway, R. N. 1989. New Zealands pre-human avifauna and its vulnerability. New Zealand Journal of Ecology 12 supplement:115-129.

IUCN. 1996. The 1996 IUCN red list of threatened animals. IUCN, Gland, Switzerland.

James, H. F. 1995. Prehistoric extinctions and ecological change in oceanic islands in Vitousek et al, ed. Ecological Studies. Springer-Verlag, Berlin.

Jenkins, M. 1992. Species extinction. In: Global biodiversity: status of the earth's living resources (B. Groombridge, ed.), Chapman & Hall, London, Pp: 192-205.

Johnson, T. H., and A. J. Stattersfield. 1990. A global review of island endemic birds. Ibis 132:167-180.

Keegan, W. F., and J. M. Diamond. 1987. Colonization of islands by humans: a biogeographical perspective. Advances in Archaeological Methods Theory 10:49-92.

King, W. B. 1985. Island birds: will the future repeat the past? International Council for Preservation, Cambridge.

Kirch, P. V. 1982. The impact of prehistoric Polynesians on the Hawaiian ecosystem. Pacific Science36:1-14.

Lawton, J.H. 1995. Population dynamic principles. In: Extinction rates (J. H. Lawton and R. M. May, eds), Oxford University Press, Oxford, Pp. 147-163.

Leigh, E. G. 1981. The average lifetime of a population in a varying environment. Journal of Theoretical Biology 90:213-239.

Lockwood, J.L, T.M. Brooks and M.L. McKinney. 2000. Taxonomic homogenization of the global avifauna.. Animal Conservation 3:27-35.

MacArthur, R.H. Geographical ecology. Harper and Row, New York.

MacArthur, R. H., and E. O. Wilson. 1967. The theory of island biogeography. Princeton University Press, Princeton.

Manne, L.L., T.M. Brooks, and S.L. Pimm 1999. Relative risk of extinction of passerine birds on continents and islands. Nature 399:258-261.

Martin, R. D. 1993. Primate origins: plugging the gaps. Nature 363:223-234.

May, R. M., J. H. Lawton, and N. E. Stork. 1995. Assessing extinction rates. In: Extinction rates (J. H. Lawton and R. M. May, eds.), Oxford University Press, Oxford, Pp: 1-24.

McCall. 1997. Biological, geographical and geological factors influencing biodiversity on islands. Unpublished D.Phil. thesis. University of Oxford, Oxford.

McLain, D. K., M. P. Moulton, and T. P. Redfearn. 1995. Sexual selection and the risk of extinction of introduced birds on oceanic islands. Oikos 74:27-34.

Milberg, P., and T. Tyrberg. 1993. Naive birds and noble savages a review of man caused prehistoric extinctions of island birds. Ecography 16:229-250.

Mitra, S., H. Landel, and S. Pruett-Jones. 1996. Species richness covaries with mating system in birds. Auk 113:544-551.

Møller, A.P. and J.J. Cuervo. 1998 Speciation and feather ornamentation in birds. Evolution 52:859-869.

Mountfort, G. 1988. Rare birds of the world. Collins, London.

Nee, S., T.G. Barraclough, and P. Harvey. 1996. Temporal changes in biodiversity: detecting patterns and identifying causes. In: Biodiversity (K.J.Gaston, ed.), Blackwell Scientific Press, Pp: 230-252.

Olson, S. L. 1977. Additional notes on subfossil bird remains from Ascension Island. Ibis 19:37-43.

Olson, S. L., and H. F. James. 1982. Fossil birds from the Hawaiian Islands: evidence for wholesale extinction by man before western contact. Science 217:633-635.

Owens, I.P.F and P.M. Bennett (submitted). Revealing the ecological correlates of extinction-risk in birds. 13 pp.

Owens, I.P.F., P.M. Bennett, and P.H. Harvey. 1999. Species richness among birds: body size, life history, sexual selection or ecology? Proceedings Of the Royal Society Of London Series B-Biological Sciences 266:933-939.

Pimm, S.L. 1991. The balance of nature. University of Chicago Press.

Pimm, S. L., H. L. Jones and J. M. Diamond. 1988. On the risk of extinction. American Naturalist 132:757-785.

Pimm, S. L., M. P. Moulton, and L. J. Justice. 1994. Bird extinctions in the central Pacific. Proceedings Of the Royal Society Of London Series B-Biological Sciences 344:27-33.

Pimm, S. L., M. P. Moulton, and L. J. Justice. 1994. Bird extinctions in the central Pacific. In: Extinction rates (J. H. Lawton and R. M. May, eds.), Oxford University Press, Oxford, Pp: 75-87.

Pimm, S. L., G. J. Russell, J. L. Gittleman, and T. M. Brooks. 1995. The future of biodiversity. Science 269:347-350.

Prestwich, A. A. 1976. Extinct, vanishing, and hypothetical Parrots. Pages 198-204. Avicult.

Raup, D.M. 1991 Extinction: bad genes or bad luck? Oxford University Press.

Roberts, M. 1991. Origin, dispersal routes, and geographic distribution of Rattus exulans, with reference to New Zealand. Pacific Scientist 45:123-130.

Rosenzweig, M.L. Species diversity in space and time. Cambridge University Press.

Russell, G. J., T. M. Brooks, M. L. McKinney, and C. G. Anderson. 1998. Change in taxonomic selectivity in the future extinction crisis. Conservation Biology 12:1365-1376

Sibley, C.G. and J.E. Ahlquist. 1990. Phylogeny and classification of birds. Yale University Press, New Haven.

Sibley, C.G. and B.L. Monroe 1990. Distribution and taxonomy of birds of the world. Yale University Press, New Haven.

Simberloff, D. 1992. Extinction, survival, and effects of birds introduced to the Mascarenes. Acta Oecologia 13:663-678.

Sorci, G., A. P. Moller and J. Clobert. 1998. Plumage dichromatism of birds predicts introduction success in New Zealand. Journal of Animal Ecology 67:263-269.

Stattersfield, A. J., M. J. Crosby, A. J. Long and D. C. Wege. 1998. Endemic bird areas of the world: priorities for biodiversity conservation. Burlington Press, Cambridge.

Steadman, D. W. 1991. Extinction of species: past, present and future. In: Global climate change and life on earth, (R. L. Wyman, ed.), Chapman & Hall, New York, Pp: 156-169.

Steadman, D. W. 1995. Prehistoric extinctions of Pacific island birds: biodiversity meets zooarchaeology. Science 267:1123-1131.

Steadman, D. W., and P. S. Martin. 1984. Extinction of the birds in the late Pleistocene of North America. In: Quaternary extinctions: a prehistoric revolution (P. S. Martin and R. G. Kline, eds.), University of Arizona Press, Tucson, Pp: 466-477.

Steadman, D. W., J. P. White, and J. Allen. 1999. Prehistoric birds from New Ireland, Papua New Guinea: extinctions on a large Melanesian island. Proceedings of the National Academy of Science 96:2563-2568.

Terborgh, J. 1974. Preservation of natural diversity: the problem of extinction prone species. BioScience 24:715-722.

Terborgh, J. and B. Winter. 1980. Some causes of extinction. In: Conservation Biology: an evolutionary-ecological perspective (M. E. Soule and B. A. Wilcox, eds.), Sinauer Associates, Sunderland, Mass., Pp: 119-133.

Wragg, G. M. 1995. The fossil birds of Henderson Island, Pitcairn Group: natural turnover and human impact, a synopsis. Biological Journal of the Linnean Society 56: 405-414.

Chapter 11

Downsizing Nature: Anthropogenic Dwarfing of Species and Ecosystems

Mark V. Lomolino[1], Rob Channell[2], David R. Perault[3] and Gregory A. Smith[4]
[1]*Oklahoma Biological Survey, Oklahoma Natural Heritage Inventory and Department of Zoology, University of Oklahoma, Norman, OK, 73019, USA*
[2]*Department of Biological Sciences, Fort Hays State University, Hays KS, 67601, USA*
[3]*Environmental Science Program, Lynchburg College, Lynchburg, VA, 24501, USA*
[4]*Department of Zoology and Oklahoma Biological Survey, University of Oklahoma, Norman, OK, 73019, USA*

1. INTRODUCTION

Few if any serious scientists question that we are witnessing an acceleration of extinction rates. The ongoing loss in biological diversity far exceeds what paleontologists consider background levels and may well, if projections are accurate, rival some of the mass extinctions of the geological record (Temple, 1986; Wilson, 1988 and 1992; World Conservation Monitoring Centre, 1992; Vitousek et al., 1997). In this case, of course, the extinctions result from the actions of one species, our own. Yet the ongoing surge in extinctions is just the most recent in a long series of anthropogenic pulses of extinctions - each following a new wave of human colonization (Figure 1), or the spread of some technical advance in our ability to dominate native landscapes (e.g., through the use of fire, water diversion, domestication of animals, mechanized agriculture, timber harvest and fisheries; see Howells, 1973; Clark, 1992; Flannery, 1994; Gamble, 1994).

Each of these successive waves in colonization and extinction differed in some key features, including their magnitude, location and spatial extent. Yet in common among them is that diversity plummeted, not just by the sheer number of species lost, but by the highly disproportionate loss of many of the most distinctive species and ecosystems. From the earliest to most advanced human societies, our colonization and domination of native biotas and landscapes has been highly selective and non-random (see Martin, 1967, 1984 and 1990;

McKinney, 1997; McKinney and Lockwood, 1999; Manne et al., 1999). We tend to advance most rapidly and have our greatest impact on coastal and other lowland ecosystems where we then select the largest packets of protein and energy (i.e., the megafauna) and exclude other species that would compete with or prey on humans and our livestock.

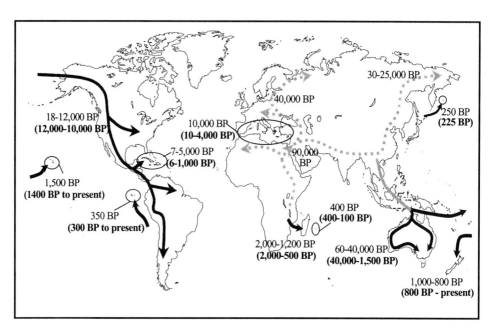

Figure 1. Colonization of the globe by anatomically modern, and ecologically significant humans (*Homo sapiens sapiens;* dates on top line) was often followed by waves of megafaunal extinctions (dates below, in bold and parentheses; sources were Gamble, 1994; MacPhee and Marx, 1997; see also Brown and Lomolino, 1998). Dashed arrows indicate earliest waves of colonization of *H. s. sapiens,* otherwise arrows are shaded from early (light gray) to most recent (black) colonizations. Colonization of Africa and Asia by *H. s. sapiens* were not followed by marked extinction events, probably because the native megafauna of these regions shared an ecological and evolutionary history with modern humans.

This highly selective process has occurred at least since the earliest megafaunal extinctions of the Pleistocene (perhaps as early as 60,000 years BP in Australia and New Guinea; see MacPhee and Marx, 1997; Flannery, 1994; Figure 1). It is what we term *anthropogenic homogenization* - the highly disproportionate loss in the most distinctive elements (both species and ecosystems) of biological diversity. Here we focus on three fundamental and related features of biological diversity: body size, ecosystem patch size and geographic range size. Although our study is restricted to empirical patterns for mammals, we think the lessons are both general and compelling. Not only have we caused the extinctions of thousands of plants and animals, but we have reversed some of nature's most general ecological and evolutionary trends.

2. BODY SIZE - MEGAFAUNAL EXTINCTIONS AND TIME DWARFING

Throughout the evolutionary history of the class Mammalia, especially during the past 65 million years, the rise in the number of species has been coupled with a diversification of their morphologies. Most tractable and perhaps most important among these morphological features is body size. As is common in most groups of plants, animals and other forms of life, the frequency distribution of body size in mammals is leptokurtic and right-skewed (Figure 2; see also Silva and Downing, 1995; Blackburn and Gaston, 1998). Mammals exhibit a tremendous range in body mass (from ~ 2 g to over 120,000 kg; Nowak, 1999), but most species are relatively small (< 1 kg; see also Brown and Nicoletto, 1991). According to Cope's Rule, body size of most groups of mammals tends to increase over evolutionary time (Cope, 1887; Brown and Maurer, 1986; Alroy, 1998; Maurer, 1998). That is, not only has the number of mammals increased, but the number of larger species has persistently increased, at least until recent times.

Whether it was the aboriginal Australians preying on diprotodons and other mega-marsupials, the early Polynesians who devastated over 15 species of moas on New Zealand, European sailors and colonists of other oceanic islands raiding dodos, solitaires, great auks and giant tortoises, or the early colonists of North America that probably caused extinctions of that continents megafauna in the late Pleistocene and early Holocene, these anthropogenic extinctions were far from random. In each case, the native fauna lost a very disproportionate number of highly distinctive (larger) species (Figure 3). Over the multiple waves of prehistoric and historic colonization, we have homogenized the body size profile of the class Mammalia, collapsing the frequency distribution of body size toward smaller and smaller sized species (Figures 2 and 3; see also Martin, 1967 and 1984; Caughley, 1987 Owen-Smith, 1987 and 1989; Flannery, 1994; Alroy, 1999). Based on the number of larger species that are now endangered or threatened, anthropogenic homogenization of morphological distinctiveness of mammals is likely to continue (Figure 2).

This reversal of Cope's Rule, one of nature's most persistent macroecological trends, is also evident at the microevolutionary level. Tim Flannery (1994) provides a fascinating and compelling case for what he terms "time dwarfing" in Australian mammals. Not only did the mammals of that continent experience a substantial and highly selective loss of its largest species following aboriginal colonization (some 40,000 to 60,000 years BP), but many of the surviving large mammals became dwarfed. By removing the largest predators, preying on larger individuals of remaining species, and modifying native landscapes by the use of fire, it appears that aborigines shifted selective pressures to favor smaller individuals of these otherwise large mammals (Figure 4). Consistent with both the micro- and macroevolutionary patterns, the magnitude of time dwarfing was greatest for the largest species (Flannery, 1994).

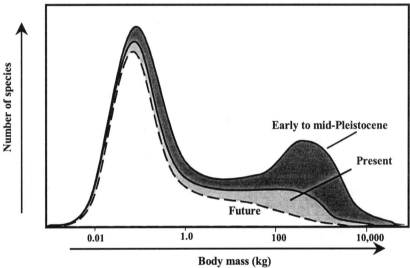

Figure 2. Downsizing of mammals is apparent in the differential loss of larger mammals during the Holocene (see Caughley, 1987), and in the highly disproportionate number of large mammals listed as imperiled. The prospective extinctions of this latter group (today's remaining, but imperiled megafauna) will likely contribute to the continued homogenization of mammalian body size.

2.1 Parable of the Woolly Mammoth

As implied above, dwarfing in mammals may result from direct (selective predation) as well as indirect effects (from removal of mega-predators and habitat modification) of human activities. The account of the decline and eventual extinction of the woolly mammoth (*Mammuthus primigenius*) is a poignant, illustrative case study. During the late Pleistocene the range of this species covered much of North America and Eurasia north of 40° N latitude (Figure 5; see Martin, 1990 and 1995;Vartanyan et al., 1993 and 1995). With the recession of the glaciers, human civilizations expanded their ranges throughout this region and began to prey more heavily on this species and other members of the continental megafauna. As a result of this accelerated predation

10. Downsizing Nature

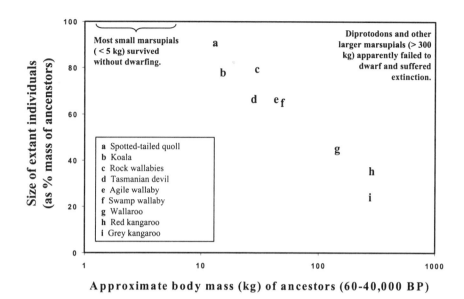

Figure 3. *Larger mammals tend to be disproportionately common in the record of Pleistocene extinctions, and among the list of imperiled species in the recent biota of Australia (after Flannery, 1994).*

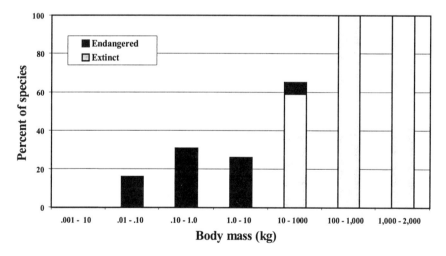

Figure 4. *Time-dwarfing in Australian mammals during the Pleistocene and early Holocene (after Flannery, 1994). Australian mammals smaller than 5 kg were little affected by aboriginal colonization of the continent. Megafaunal mammals suffered extinctions, while a number of medium-sized mammals exhibited a graded trend of dwarfism.*

from a novel predator (perhaps along with concurrent climate change), most of the megafaunal species of North America suffered extinctions (Martin, 1967 and 1984; Alroy, 1999). The mammoths, although extirpated from the mainland portions of their range by around 11,000 BP, survived on Wrangel Island (north

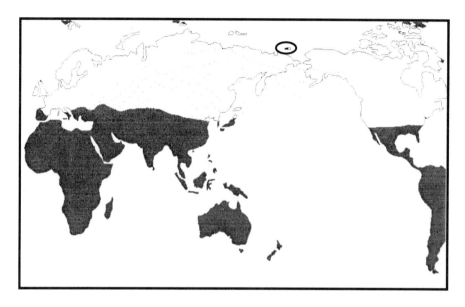

Figure 5. Patterns of range collapse in the Woolly mammoth (*Mammuthus primigenius*). Although this species was extirpated from the continental landmasses around 10,000 BP, a relictual population of dramatically dwarfed individuals persisted on Wrangel Island (circled) until about 3,700 BP (Vartanyan et al., 1993; geographic range and location of ice sheets during last glacial maximum in light gray and white, respectively; final range on Wrangel Island in black).

of Siberia) well into the Holocene (Vartanyan et al., 1993 and 1995). Now isolated from the mainland and its large predators, including humans, surviving mammoths underwent very rapid and truly remarkable dwarfism (see also Lister [1989] for a general account of rapid dwarfism in Pleistocene elephants). Body mass of Wrangle Island mammoths was just a third that of their ancestors. These dwarfed mammoths (now a true oxymoron) went extinct around 3 to 4000 BP, some 10,000 year after the glaciers receded, just after apparent colonization of Wrangel Island by humans.There remains some question over whether megafaunal extinctions were caused by human populations, climate change, or a combination of these global scale events (e.g., see Martin, 1984; Steadman and Martin, 1984; Graham, 1986; Owen-Smith, 1987; Barnosky, 1989; Beck, 1996;

Macphee and Marx, 1997). However, it is becoming clear that each successive wave of colonization by ecologically significant humans was followed by a wave of extinctions which removed a highly disproportionate number of large species (Figure 1; see also Martin, 1984; MacPhee and Marx, 1997). The reoccurrence of these coupled events, human colonization and megafaunal extinctions, across most of the globe and at very different time periods argues strongly in favor of anthropogenic versus climatic causes for megafaunal extinctions (but see also Graham, 1986). Regardless of the ultimate outcome of this debate on prehistoric extinctions, time dwarfing and extinction of insular mammoths serves as an important parable for conservation biology. Modern and future human societies will continue to have a significant and highly selective impact on native biotas, causing disproportionate loss of large sized and large-ranging species, and causing dwarfing and other morphological, physiological and genetic changes in many surviving species.

3. DOWNSIZING NATIVE ECOSYSTEMS

Analogous to anthropogenic changes in body size discussed above, human activities have reduced the variance in patch size of native ecosystems, largely through a highly disproportionate loss or fragmentation of the largest patches. Our domination of native landscapes is so pervasive that it is difficult to find any ecosystem that has not been significantly reduced, fragmented or otherwise altered by humans (see Vitousek et al., 1997). Archaeological evidence from sites as distant as Easter Island, the ancient lands of the Incas, and western Europe during the Middle Ages indicates that the ability to transform native landscapes is not unique to modern societies (Fagan, 1990; Guilaine, 1991; Flannery, 1994; Gamble, 1994). The key distinction between the ecological powers of prehistoric and modern societies is that now there are many more of us and that we now act on a global scale.

Unlike the loss in species diversity, it is difficult to document the total destruction of any particular ecosystem at the hands of humanity. Yet many natural ecosystems have been significantly altered - transformed from once expansive stands of forests, prairies and other habitats to archipelagoes of ever dwindling habitat islands (Harris, 1984; Wilcox and Murphy, 1985; Wilcove, 1987; Wilson, 1988; Saunders et al., 1991). Here we provide just two examples of downsizing of native ecosystems, both stemming from our field studies of mammals inhabiting two landscapes of North America. We believe the emergent patterns of ecosystem dynamics are representative of those for a very broad range of natural ecosystems and landscapes.

3.1 Prairie Dog Towns of the Great Plains Region, North America

During the late 1800s, much of central North America (between 40 and 100 million ha) was covered with prairie dog towns (Miller et al., 1994; Figure 6). Of the five species of prairie dogs in this region, the black-tailed prairie dog (*Cynomys ludovicianus*) had by far the largest geographic range. By the turn of the century, however, the U.S. Biological Survey (led by their new director - C. H. Merriam), declared war on prairie dogs. Through a combination of poisoning and shooting campaigns, land conversion and repeated occurrence of the exotic plague, the range and population numbers of prairie dogs were reduced to just 2 % of historic levels (Miller et al., 1994). Towns of over 1000 ha may have been commonplace in the 18th and 19th century. One of the largest towns stretched across 22 miles of central Oklahoma prior to control programs. Assuming for the sake of simplicity that this town was circular, its area can be estimated at over 98,000 ha. Black-tailed prairie dog towns have undergone marked "dwarfing" during the 20th century. By the 1980s mean town size in Oklahoma was reduced to just 18 ha, and has now fallen to less than 11 ha (Figure 6; see Lomolino and Smith, 2000).

While it may well be impossible to identify all the direct and indirect effects of decimation and downsizing of these prairie ecosystems, the ongoing and imminent loss in biological diversity of native communities is obvious. The largest town specialist, the black-footed ferret (*Mustela nigripes*) is now extirpated from all but a tiny fraction of its former range persisting largely by the yeoman efforts of biologists guiding the protective management and captive breeding of this species (Seal et al., 1989; Anderson et al., 1986). The declines of other predators and species closely associated with burrows and mowed habitats maintained by prairie dogs seem well-documented (e.g., swift fox [*Vulpes velox*], thirteen-lined ground squirrel [*Spermophilus tridecemlineautus*], grasshopper mouse [*Onychomys leucogaster*], burrowing owl [*Speotyto cunicularia*] , golden eagles [*Aquila chrysaetos*], ferruginous hawks [*Buteo regalis*], horned lark [*Eremophila alpestris*], mountain plover [*Charadrius montanus Charadrius montanus*], prairie rattlesnake [*Crotalus viridis*], Texas horned lizard [*Phrynosoma cornutum*], tiger salamanders [*Ambystoma tigrinum*], spadefoot toads [*Scaphiopus spp.*], spiders and many other invertebrates; Miller et al., 1994; Kotliar et al., 1999; personal observations, M. V. Lomolino and G. A. Smith). We predict, therefore, that smaller towns will harbor both fewer and a highly non-random subset of native species: i.e., those species pre-adapted to relatively small and isolated towns.

Which species are associated with prairie dog towns should depend on the particular region and pool of species. That is, town associates may vary substantially among ecoregions. Therefore, along with the numeric and spatial

10. Downsizing Nature

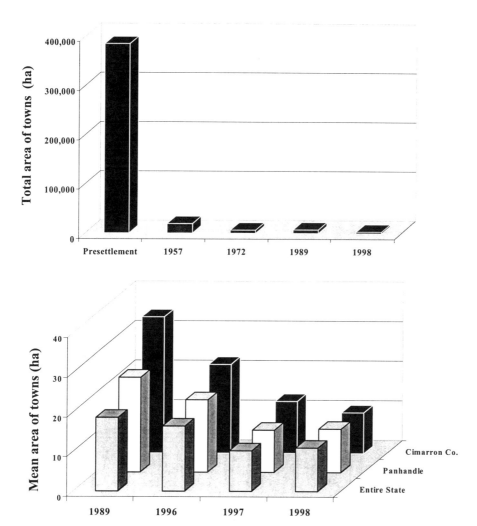

Figure 6. Geographic dynamics of black-tailed prairie dog towns along the south-eastern edge of their historic range (i.e., Oklahoma, USA) also exhibit downsizing, or the simultaneous and significant decline in both total and mean area of remaining towns. Cimarron County is the western-most county in the Panhandle (northwestern) region of Oklahoma.

decline in prairie dog towns across their entire range, we should be equally concerned with the coincident decline in frequency distribution of towns among biogeographic and ecological regions. For example, the loss in biological diversity associated with prairie dog towns may be much higher in remnant stands of mixed grass prairie and other ecoregions that are subject to intense agricultural pressures.

3.2 Temperate Rainforests of the Pacific Northwest

Despite the obvious differences between native prairies and old-growth forest ecosystems, they are both subject to anthropogenic decline in coverage and concomitant fragmentation and downsizing of remnant patches. Rainforests are of special concern to conservation biologists because they represent hotspots of biological diversity and of ongoing habitat loss. Both scientists and lay people alike are familiar with the great diversity of tropical rainforests, where communities that cover just 7% of earth's land surface harbor over half of its species. Therefore, deforestation of tropical rainforests, estimated at roughly 76,000 km^2 per yr (World Conservation Monitoring Centre, 1992), is one of the most serious threats to biological diversity on a global scale.

In contrast to their tropical counterparts, temperate rainforests cover much less area (about 5 vs. 17 million km^2; Whittaker and Likens, 1973), are much less diverse in total number of plants and animal species, but they house some of the world's most ancient communities and it's largest life forms (see Norse, 1990). Common within these old-growth rainforests are trees well over 80 m tall and 200 years old. Unfortunately, such stands of titan trees also hold enormous economic value and have been subject to heavy logging and clearcutting, especially during the past five or six decades. Timber harvest in areas such as the Pacific Northwest of North America followed the very common pattern for humans to first colonize and dominate the lowland communities. Thus, prior to the 1950s, most timber harvest in the Olympic Peninsula was limited to elevations below 500 m (Figure 7a). With technological advances of the timber industry, along with depletion of readily accessible forests of the lowlands, timber harvest began to climb to higher elevations.

The resultant pattern of landscape transformation should now sound familiar. The expansive and essentially continuous rainforests of the Olympic Peninsula were rapidly converted to an archipelago of dwindling and more isolated fragments of old-growth forests. Overall reduction and fragmentation of remnant stands continued until a partial moratorium of timber harvest was declared in the 1980s. Yet, by this time, the Olympic Peninsula's old-growth had been reduced to less than 20% of its historic coverage (Morrison, 1989) and mean fragment size in our study area declined from 1267 ha in 1950, to just 166 ha in 1990 (Figure 7a, b; Perault, 1998; Lomolino and Perault, 2000a and b; Perault and Lomolino, 2000).

Again, the predicted ecological effects of downsizing of these ancient forests is a decline in biological diversity, especially for the relatively larger, broad-ranging specialists and top carnivores. In fact, the decline in populations of predators including spotted owls, fishers and martens is well-documented (Forsman et al., 1984; Raphael, 1988; Norse, 1990). In addition to the significant effects on wildlife communities in this region, reduction in size of forest fragments appears to have had significant effects at the intraspecific level as well. For example, demographic and behavioral characteristics of spotted owls inhabiting relatively small fragments (those < 50 ha) differ significantly from those occupying larger and relatively continuous stands of old-growth forests (Carey et al., 1992).

In our own studies on the effects of fragmentation and habitat reduction on mammals of the Olympic National Forest (Songer et al., 1997; Lomolino and Perault, 2000a, b; Perault and Lomolino, 2000), we found that small fragments of old-growth forest (those < 50 ha) are dominated by broad-ranging, matrix species including spotted skunks (*Spilogale putorius*), black-tailed deer (*Odocoileus hemionus*), and the common deer mouse (*Peromyscus maniculatus*). Absent or rare in these small remnants of old-growth are such native, old-growth species as southern flying squirrels (*Glaucomys sabrinus*), red-backed voles (*Clethrionomys gapperi*), forest deer mice (*Peromyscus oreas*) and Trowbridge and montane shrews (*Sorex trowbridgii* and *S. monticolus;* Songer et al., 1997; Lomolino and Perault, 2000a, b; Perault and Lomolino, 2000). Thus, many geographically restricted, old-growth specialists are being replaced by a few, cosmopolitan generalists - species pre-adapted to human-dominated systems. Finally, we note that despite the great differences in two native ecosystems - temperate rainforests and prairie dog towns, their biogeographic dynamics are quite similar. Both of these native ecosystems are collapsing in size, with the remnant patches becoming increasingly more isolated.

3.3 Geographic Range Dynamics

One of the most general patterns emerging from the relatively new field of macroecology (see Brown, 1995) is the relationship between geographic range size and body size. The emergent pattern, one that appears to have been anticipated by Charles Darwin, is basically this. For mammals, birds and most other groups of animals, geographic range size varies within the bounded space set by two constraints (Figure 8). First, regardless of the body size of the focal species, maximum geographic range size is constrained by principle features of

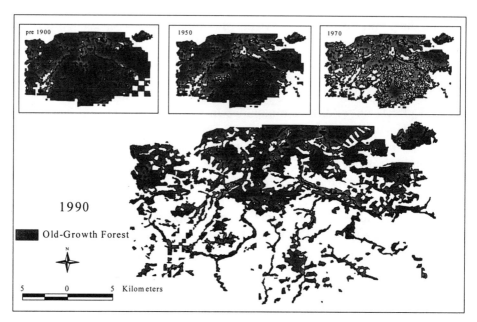

Figure 7a. Fragmentation of old-growth, temperate rainforests across the Hood Canal District, Olympic National Forest, Washington, USA.

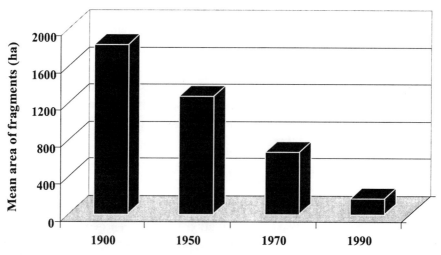

Figure 7b. Downsizing of fragmented, temperate rainforests across the Hood Canal District, Olympic National Forest, Washington, USA.

10. Downsizing Nature 235

the landmass (e.g., geographic barriers and total area of the continent). Second, minimum geographic range size varies directly with body size of the focal species. Thus, while range size of small species varies substantially between the maximum and minimum values of the entire species group, species of larger body size have larger (and less variable) geographic ranges. There are of course exceptions, albeit transient ones, to this pattern - namely, the large species whose geographic ranges are collapsing as they decline towards extinction (Figure 8, dashed arrow).

Based on the anthropogenic patterns discussed in previous sections, we predict that the very general, macroecological pattern illustrated in Figure 8 will change substantially over the coming decades. Not only will some species be lost from all areas of the bounded space, but a highly disproportionate number of large bodied, broad-ranging species will be lost. Whether large or small, species whose ranges fall near the lower constraint line (i.e., those whose ranges are marginal relative to their body size) are geographically imperiled. Such species may undergo time dwarfing, remain imperiled, or their geographic ranges will continue to collapse until they become extinct. The ultimate effect of downsizing native biotas and their habitats will be a significant loss in biological and biogeographic diversity, with an upward ratcheting of the lower constraint line of this macroecological pattern (Figure 8). A few, matrix adapted species (e.g., skunks, deer, coyotes and common deer mice) will expand their ranges at the expense of the native megafauna, specialists and geographically restricted species.

Finally, some recent studies provide additional insights on patterns of geographic range collapse in endangered species. These patterns are consistent with the hypothesis that geographic range decline is largely determined by the spatial dynamics of humans and their commensals (see Lomolino and Channell, 1995 and 1998; Channell, 1998; Channell and Lomolino, 2000a and 2000b; see also Maurer et al., 2000; Brooks, 2000). Because human civilizations and technological advances tend to spread across the landscape much like a contagion, species populations decline in a similar fashion. Initial extinctions tend to involve a peripheral population at the edge of the expanding extinction front (i.e., those first to be overwhelmed by the spread of human populations, deforestation, exotic predators, competitors or disease). Eventually, however, the extinction front spreads to include populations at the core of the species' historic range, leaving just a few populations on an island, mountaintop or otherwise isolated, peripheral area of the species' geographic range (Figure 9; see Bibby, 1994; Short and Smith, 1994; Towns and Daugherty, 1994).

Just as important, these studies on geographic range collapse also reveal that, rather than simply declining as one continuous range, geographic ranges tend to fragment into a number of small ranges well before the species becomes extinct

(Channell, 1998; see also Maurer et al., 2000). Now reconsider this process within the framework of the macroecological pattern of Figure 8. In contrast to what we would expect for the decline of one, continuous geographic range, ranges of each of these now isolated and fragmented populations will fall much closer to the lower constraint line: i.e., the line associated with geographically imperiled species across a wide range of body size (Figure 8).

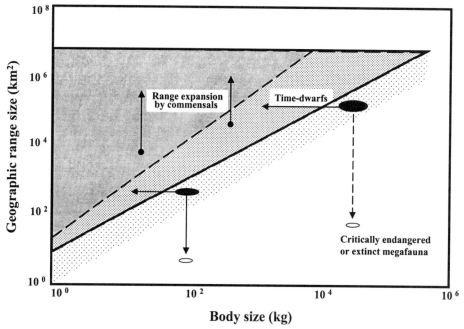

Figure 8. Geographic range size in mammals tends to vary within the constrained space delimited by the solid lines in this figure. While the upper constraint line probably depicts physiographic limits on this biotic group (i.e., it is a function of size of the land mass), the lower (diagonal) constraint line illustrates how minimal range requirements vary with body size of the species. Species within this zone of geographic imperilment (below diagonal line) will either expand their range, providing they are pre-adapted to anthropogenic habitats, undergo time dwarfing, or suffer continued range collapse and eventual extinction.

4. CONCLUSION

We now return to the parable of the woolly mammoth - a once enormous, ecologically dominant species that ranged across a vast geographic range during the Pleistocene. This species found its final refuge, albeit in a markedly transformed body plan, on an island isolated from the early waves of ecologically significant humans that colonized lands at the heels of the receding glaciers. Today we find that islands, and island-like montane ecosystems and nature reserves now serve as final refugia for a great diversity of species (Bibby, 1994; Short and Smith, 1994; Towns and Daugherty, 1994). We fear that many of these species may ultimately suffer extinction, or persist in a highly altered form.

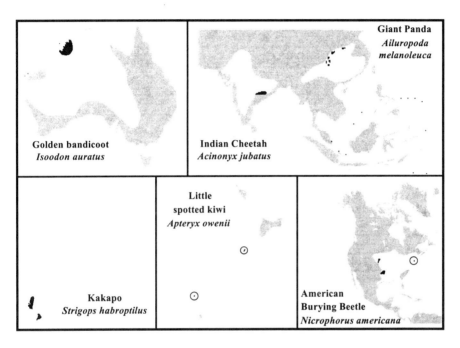

Figure 9. Patterns of geographic range collapse in six species of animals (historic and current geographic ranges in light gray and black, respectively; Lomolino and Channell, 1995; Channell and Lomolino, 2000a and 2000b).

We have only to look to our modern "mammoths", African and Asian elephants (*Loxodonta africana* and *Elaphus maximus*), for an apparent case of persistence in an altered form. Pleistocene elephants (mastodons, mammoths and other Proboscideans) are just a few of the many mammals that conform to what has been termed the island rule: on islands, larger species of mammals tend to undergo dwarfism while smaller species tend to increase in body size (Foster, 1963 and 64; Lomolino, 1985; Heaney, 1978). This pattern is evidenced in both contemporary and prehistoric mammals (Figure 10). Dwarfing in Pleistocene elephants, including those of Wrangel Island, the Channel Islands off the coast of California, and islands in the Mediterranean Sea, has been attributed to resource limitation, adaptations for a more rugged terrain, and the absence or paucity of large predators (Hooijer, 1967; Sondaar, 1977; Johnson, 1978; Lister, 1993; Roth, 1996), which presumably preyed most heavily on smaller elephants.

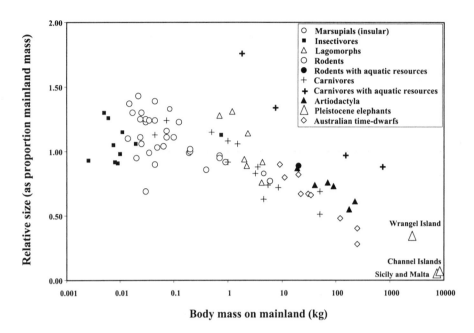

Figure 10. The island rule describes the graded trend from gigantism in insular populations of small mammals to dwarfism in insular populations of otherwise large mammals. The degree of dwarfism exhibited by Australian time dwarfs following colonization by aborigines, and that exhibited by insular elephants of the Pleistocene is consistent with the pattern for extant mammals (see Lomolino, 1985; Flannery, 1994).

In a provocative manuscript, Joseph Dudley (2000) warns that this island scenario may be recurring in modern times, but not just on true islands. Remnant populations of both African and Asian elephants are now restricted to island-like nature reserves, many of which lack mega-carnivores (e.g., tigers and lions). Dudley predicts that unless populations of these mega-carnivores are restored (or emulated by wildlife managers), elephants of tomorrow's reserves may differ markedly from their ancestral populations. In the absence of culling of smaller individuals by mega-carnivores, elephant body size may decline substantially. Isolated populations of elephants have already undergone dramatic changes in another morphological feature - their tusks. Ivory poaching, although uncontrolled by definition, is far from random, with poachers preying most heavily on individuals with the largest tusks (which often means the largest individuals, as well). As a direct result, the incidence of tusklessness has increased substantially in many elephant populations, becoming genetically fixed in some (Shoshani and Eisenberg, 1982; Sukumar, 1992; Abe, 1996).

We can only speculate on the possible generality of such intraspecific shifts and homogenization of other species now persisting in archipelagos of isolated and dwindling reserves. There is some limited evidence suggesting that fragmentation and downsizing of native ecosystems has led to decreased size and other morphological changes in other species (e.g., the prickly forest skink [*Gnypetoscinus queenslandiae*] an endemic of tropical rainforests of Queensland Australia [Summer et al., 1999]; African rainforest birds [Smith et al., 1997]; and a cyprinid fish [*Lavinia symmetricus*] of degraded freshwater in California, USA [Brown et al., 1992]). As we mentioned earlier when discussing behavioral and demographic characteristics of spotted owls, intraspecific responses to downsizing need not be restricted to changes in morphology.

The Wrangel Island mammoth was an evolutionary marvel, but its ultimate fate is perhaps the most important lesson of this parable. The dwarfed insular mammoths were able to persist for many centuries, indeed millennia, after their mainland ancestors disappeared. But they were not the same animals that thundered across the North American and Eurasian mainlands during the Pleistocene. When these and many other members of the mainland megafauna perished, the fates of a diversity of predators and scavengers were also sealed. African and Asian elephants may persist long into the next century, along with an unknown subset of today's imperiled wildlife and plants. Yet, unless we can advance by quantum leaps in our skills as ecological and evolutionary engineers, the surviving remnants of native biotas are likely to be just dim shadows of their once marvelous and diverse ancestors.

ACKNOWLEDGEMENTS

The studies discussed here were supported by grants from the National Science Foundation, USA (DEB-9322699, DEB-9707204, DEB-9622137, DEB-9820439, and DEB-9942014), Faculty Enrichment grants from the College of Arts and Sciences of the University of Oklahoma, and a grant from the George M. Sutton Foundation. GIS data for the Olympic National Forest were supplied by William Wettengel. We are grateful to the Department of Zoology, University of Oklahoma and to the Oklahoma Biological Survey for logistical support during our research. We have benefited from numerous discussions with other ecologists and evolutionary biologists including James H. Brown, Brian A. Maurer, Michael Kaspari, Douglas Kelt, Paul S. Martin, Dov F. Sax, Morgan Ernest, Michael Rosenzweig and Alex Baynes.

REFERENCES

Abe, E. 1996. Tusklessness among the Queen Elizabeth National Park elephants, Uganda. Pachyderm 22:46-47.

Alroy, J. 1998. Cope's Rule and the dynamics of body size mass in North American fossil mammals. Science 280:731-734.

Alroy, J. 1999. Putting North America's end-Pleistocene megafaunal extinction in context: large-scale analyses of spatial patterns, extinction rates, and size distributions. Pp. 105-143 in, Extinctions in near time (R. MacPhee, ed.). Academic/Plenum Publishers, New York.

Anderson, E., S. C. Forrest, T. W. Clark, and L. Richardson. 1986. Paleobiology, biogeography, and systematics of the black-footed ferret, *Mustela nigripes* (Audubon and Bachman), 1851. Great Basin Naturalist Memoirs, 8:11-62.

Barnosky, A. D. 1989. The late Pleistocene event as a paradigm for widespread mammal extinction. Pp. 235-254, in, Mass Extinctions, S. K. Donovan, ed. Columbia University Press, New York.

Beck, M. W. 1996. On discerning the cause of late Pleistocene megafaunal extinctions. Paleobiology 22:91-103.

Bibby, C. J. 1994. Recent past and future extinctions in birds. Philosophical Transactions of the Royal Society of London, Series B., 344:35-40.

Blackburn, T., and K. J. Gaston. 1998. The distribution of mammal body masses. Diversity and Distributions 4:121-133.

Brooks, T. 2000. Are unsuccessful avian invaders rarer in their native range than successful invaders? Chapter 7 in, Biotic Homogenization: the loss of diversity through invasion and extinction (eds. M. L. McKinney and J. L. Lockwood).

Brown, J. H. 1995. Macroecology. University of Chicago Press, Chicago.

Brown, J. H. and M. V. Lomolino. 1998. Biogeography. Second ed. Sinauer Associates, Inc., Publishers, Sunderland, Massachusetts.

Brown, J. H., and B. A. Maurer. 1986. Body size, ecological dominance, and Cope's Rule. Nature 324: 248-250.

Brown, J. H. and P. F. Nicoletto. 1991. Spatial scaling of species assemblages: Body masses of North American land mammals. American Naturalist 138: 1478–1512.

Brown, L. R., P. B. Moyle, and W. A. Bennett. 1992. Implications of morphological variation among populations of California roach (*Lavinia symmetricus*) (Cypriniidae) for conservation policy. Biological Conservation 62:1-10.

Carey, A. B., S. P. Horton, and B. L. Biswell. 1992. Northern spotted owls: influence of prey base and landscape character. Ecological Monographs 62:223-250.

Caughley, G. 1987. The distribution of eutherian body weights. Oecologia 74: 319–320.

Channell, R. 1998. A geography of extinction: patterns in the contraction of geographic ranges. Ph.D. dissertation, University of Oklahoma, Norman, OK.

Channell, R., and M. V. Lomolino. 2000a. Dynamic biogeography and conservation of endangered species. Nature 403:84-86.

Channell, R., and M. V. Lomolino. 2000b. Trajectories toward extinction: dynamics of geographic range collapse. Journal of Biogeography.

Clark, G. 1992. Space, Time and Man: A Prehistorian's View. New York: Cambridge University Press.

Cope, E. D. 1887. The origin of the fittest. Appleton, New York.

Dudley, J. P. 2000 Conservation implications of evolutionary trends in insular elephant populations. ms in review.

Fagan, B. M. 1990. The Journey from Eden: The Peopling of Our World. London: Thames and Hudson.

10. Downsizing Nature

Flannery, T. F. 1994. The Future Eaters: An Ecological History of the Australasian Lands and People. Kew, Victoria, N.S.W., Australia: Reed International Books.

Forsman, E. D., E. C. Meslow, and H. M. Wright. 1984. Distribution and biology of the spotted owl in Oregon. Wildlife Monographs 87:1-64.

Foster, J. B. 1963. The Evolution of Native Land Mammals of the Queen Charlotte Islands and the Problem of Insularity. Ph.D. dissertation, University of British Columbia, Vancouver.

Foster, J. B. 1964. Evolution of mammals on islands. Nature 202: 234–235.

Gamble, C. 1994. Timewalkers: The Prehistory of Global Colonization. Cambridge, MA: Harvard University Press.

Graham, R. W. 1986. Response of mammalian communities to environmental changes during the Late Quaternary. In J. M. Diamond and T. J. Case (eds.), Community Ecology, 300–313. New York: Harper and Row.

Guilaine, J. 1991. Prehistory: The World of Early Man. New York: Facts on File Publishers.

Harris, L. D. 1984. The fragmented forest: island biogeography theory and the preservation of biotic diversity. University of Chicago Press, Illinois.

Heaney, L. R. 1978. Island area and body size of insular mammals: Evidence from the tri-colored squirrel (*Calliosciurus prevosti*) of Southwest Africa. Evolution 32: 29–44.

Hooijer, D. A. 1967. Indo-Australian insular elephants. Genetica 38:143-162.

Howells, W. 1973. The Pacific Islanders. Weidenfeld and Nicolson, London.

Johnson, D. L. 1978. The origin of island mammoths and the Quaternary land bridge history of the Northern Channel Islands. Quaternary Research 10:204-225.

Kotliar, N. B., B. W. Baker, A. D. Whicker and G. Plumb. 1999. A critical review of assumptions about the prairie dog as a keystone species. Environmental Management 24:177-192.

Lister, A. M. 1989. Rapid dwarfing of red deer on Jersey in the last interglacial. Nature 342: 539–542.

Lister, A. 1993. Mammoths in miniature. Nature 362:288-289.

Lomolino, M. V. 1985. Body size of mammals on islands: the island rule re-examined. American Naturalist 125: 310–316.

Lomolino, M. V., and R. Channell. 1995. Splendid isolation: patterns of range collapse in endangered mammals. Journal of Mammalogy 76:335-347.

Lomolino, M. V., and R. Channell. 1998. Range collapse, reintroductions and biogeographic guidelines for conservation: a cautionary note. Conservation Biology 12:481-484.

Lomolino, M. V., and D. R. Perault. 2000a. Assembly and dis-assembly of mammal communities in a fragmented temperate rainforest. Ecology (in press).

Lomolino, M. V., and D. R. Perault. 2000b. Island biogeography and landscape ecology of mammals inhabiting fragmented, temperate rainforests. (in review, Global Ecology and Biogeography).

Lomolino, M. V. and G. A. Smith. 2000. Biogeographic dynamics of prairie dog towns in Oklahoma. Special Feature of Journal of Mammalogy (in review).

MacPhee, R. D. E. and P. A. Marx. 1997. The 40,000-year plague: Humans, hyperdisease, and first-contact extinctions. In S. Goodman and B. D. Patterson (eds.), Human Impact and Natural Change in Madagascar, 169–217. Washington, DC: Smithsonian Institution Press.

Manne, L. L., T. M. Brooks and S. L. Pimm. 1999. Relative risk of extinction of passerine birds on continents and islands. Nature 399:258-261.

Martin, P. S. 1967. Prehistoric overkill. In P. S. Martin and H. E. Wright Jr. (eds.), Pleistocene Extinctions: The Search for a Cause, 75–120. New Haven, CT: Yale University Press.

Martin, P. S. 1984. Prehistoric overkill. In P. S. Martin and R. G. Klein (eds.), Quaternary Extinctions: A Prehistoric Revolution, 354–403. Tucson: University of Arizona Press.

Martin, P. S. 1990. 40,000 years of extinction on the "planet of doom." Palaeogeography, Palaeoclimatology, Palaeoecology 82: 187–201.

Martin, P. S. 1995. Mammoth extinction: Two continents and Wrangel Island. Radiocarbon 37: 1–6.

Maurer, B. A. 1998. The evolution of body size in birds. I. Evidence for non-random diversification. Evolutionary Ecology 12:925-934.

Maurer, B. A., E. T. Linder and D. Gammon. 2000. A geographical perspective on the biotic homogenization process: implications from the marcoecology of North American birds. Chapter 8 in, Biotic Homogenization: the loss of diversity through invasion and extinction (eds. M. L. McKinney and J. L. Lockwood).

Mckinney, M. L. 1997. Extinction vulnerability and selectivity: combining ecological and paleontological views. Annual Reviews of Ecology and Systematics 7:1-5.

McKinney, M. L., and J. L. Lockwood. 1999. Biotic homogenization: a few winners replacing many losers in the next mass extinction. Trends in Ecology and Evolution 14:450-.

Miller, B., G. Ceballos and R. Reading. 1994. The prairie dog and biotic diversity. Conservation Biology 8:677-681.

Morrison, P. H. 1989. Ancient forests on the Olympic National Forest: analysis from a historical and landscape perspective. Wilderness Society, Washington, D.C., USA.

Norse, E. A. 1990. Ancient forests of the Pacific Northwest. Island Press, Washington, D.C.

Nowak, R. N. 1999. Walker's mammals of the world. 6th edition. John Hopkins, New York. 1936 pp.

Owen-Smith, N. 1987. Pleistocene extinctions: The pivotal role of megaherbivores. Paleobiology 13: 351–362.

Owen-Smith, N. 1989. Megafaunal extinctions: the conservation message from 11,000 years B.P. Conservation Biology 3: 405–412.

Perault, D. R. 1998. Landscape heterogeneity and the role of corridors in determining the spatial structure of insular mammal populations . Ph. D. Dissertation, University of Oklahoma, Norman, Oklahoma, USA.

Perault, D., and M. V. Lomolino. 2000. Corridors and Mammal Community Structure across a Fragmented, Old-growth Forest Landscape Ecological Monographs (in press).

Raphael, M. G. 1988. Long-term trends in abundance of amphibians, reptiles, and mammals in Douglas-fir forests of northwestern California. Pages 23-31 in R. C. Szaro, K. E. Severson, and D. R. Patton (technical coordinators). Proceedings of the symposium management of amphibians, reptiles, and small mammals in North America. USDA Forest Service, General Technical Report RM-166.

Roth, V. L. 1996. Pleistocene dwarf elephants of the California Islands. Pp. 249-253, in, The Proboscidea: evolution and paleoecology of elephants and their relatives. J. Shoshani and P. Tassi, eds. Oxford University Press, Oxford.

Saunders, D. A., R. J. Hobbs and C. R. Margules. 1991. Biological consequences of ecosystem fragmentation: A review. Conservation Biology 5: 18–32.

Seal, U. S., E. T. Thorne, M. A. Bogan and S. H. Anderson. 1989. Conservation biology and the black-footed ferret. Yale University Press, New Haven. 302 pp.

Short, J., and A. Smith. Mammal decline and recovery in Australia. Journal of Mammalogy 75:288-297.

Shoshani, J., and J. F. Eisenberg. 1982. *Elaphus maximus*. Mammalian Species (American Society of Mammalogists) 182:1-8.

Silva, M, and J. A. Downing. 1995. CRC handbook of mammalian body masses. CRC Press, Boca Raton.

Smith, T. B., R. K. Wayne, D. J. Girman, and M. W. Bruford. 1997. A role for ecotones in generating rainforest biodiversity. Science 276:1855-1857.

Songer, M., Lomolino, M. V., and D. Perault. 1997. Habitat selection and niche dynamics of deer mice (*Peromyscus oreas* and *P. maniculatus*) in a fragmented, old-growth forest landscape. Journal of Mammalogy (Special Feature) 78:1027-1039.

Sondaar, P. Y. 1977. Insularity and its effect on mammal evolution. Pp. 671-703, in, Major patterns in vertebrate evolution. M. K. Hecht, P. C. Goody and B. M. Hecht, eds. Plenum Press, New York.

Steadman, D. W. and P. S. Martin. 1984. Extinctions of birds in the late Pleistocene of North America. In P. S. Martin and R. J. Klein (eds.), Quaternary Extinctions, 466–477. University of Arizona Press, Tucson.

Sukumar, R. 1992. The Asian elephant: ecology and management. Cambridge University Press, Cambridge.

Summer, J., C. Moritz and R. Shine. 1999. Shrinking forest shrinks skinks: morphological change in response to rainforest fragmentation in the prickly forest skink (*Gnypetoscinus queenslandiae*). Biological Conservation 91:159-167.

Temple, S. 1986. The problem of avian extinctions. Current Ornithology 3: 453–485.

Towns, D. R., and C. H. Daugherty. 1994. Patterns of range contractions and extinctions in the New Zealand herpetofauna following human colonisation. New Zealand Journal of Zoology 21:325-339.

Vartanyan, S. L., V. E. Garutt and A. V. Sherr. 1993. Holocene dwarf mammoths from Wrangel Island in the Siberian Arctic. Nature 362: 337–340.

Vartanyan, S. L., K. A. Arslanov, T. V. Tertychnaya and S. B. Chernov. 1995. Radiocarbon dating evidence for mammoths on Wrangel Island, Arctic Ocean, until 2000 B.C. Radiocarbon 37: 1–6.

Vitousek, P. M., H. A. Mooney, J. Lubchenco and J. M. Melillo. 1997. Human domination of earth's ecosystems. Science 277: 494–499.

Whittaker, R. H., and G. E. Likens. 1973. Primary production: the biosphere and man. Human Ecology 1:357-369.

Wilcove, D. S. 1987. From fragmentation to extinction. Natural Areas Journal 7: 23–29.

Wilcox, B. A. and D. D. Murphy. 1985. Conservation strategy: the effects of fragmentation on extinction. American Naturalist 125: 879–887.

Wilson, E. O. (ed.) 1988. Biodiversity. Washington, DC: National Academy Press.

Wilson, E. O. 1992. The Diversity of Life. Cambridge, MA: Belknap Press.

World Conservation Monitoring Centre. 1992. Global Biodiversity: Status of the Earth's Living Resources. Chapman and Hall, New York.

Chapter 12

Spatial Homogenization of the Aquatic Fauna of Tennessee: Extinction and Invasion Following Land Use Change and Habitat Alteration

Jeffrey R. Duncan[1] and Julie L. Lockwood[2]
[1]*Department of Ecology and Evolutionary Biology and Water Resources Research Center, 600 Henley St., University of Tennessee, Knoxville, TN 37996, USA*
[2]*Department of Environmental Studies, Natural Sciences II, University of California, Santa Cruz, CA 95064, USA*

1. INTRODUCTION

One of the most profound legacies of modern society is the global transformation of the world's flora and fauna due to biotic homogenization (McKinney and Lockwood 1999, Lockwood et al. 2000). The systematic loss of vulnerable taxa to extinction is a common theme in our discussions of aquatic ecosystems (Allan and Flecker 1993). Similarly, increasing rates of invasion of fresh- and saltwater communities by non-native species continue to invoke social, economic, and ecological consequences (Carlton 1996, 1992). Together, the sum of simultaneous human-facilitated extinction and invasion has resulted in the non-random homogenization of aquatic faunas worldwide. Certain higher taxa (e.g., families) may be more likely to become extinct or to invade novel territories than would be predicted by chance (Angermeier 1995, McKinney 1998, Lockwood 1999). This non-random pruning and geographic rearranging of the aquatic and marine evolutionary tree has resulted in the loss of diversity among higher taxa as well as a loss of global species richness (McKinney and Lockwood 1999).

An equally important component to biotic homogenization is the decline of diversity among historically distinct faunas (Harrison 1993). Historically, restriction of species ranges has resulted from a host of factors including physical isolation (e.g., mountain ranges), evolutionary isolation (e.g., speciation, co-evolution), and ecological isolation (e.g., inter-specific

Biotic Homogenization, edited by J. Lockwood and M. McKinney
Kluwer Academic/Plenum Publishers, 2001

competition, habitat requirements) (Huston 1994). These isolating pressures inevitably produce regionally distinct communities. In the face of habitat alterations, land use changes, and taxonomic homogenization, the effects of isolating pressures have lessened such that formerly distinct regional communities tend to resemble one another more so than if the isolating boundaries were still functional (Lodge 1993). If one combines the loss of isolation with the loss of endemic species, ecological similarity between formerly unique regions increases. Here, we explore the fate of the freshwater fish, mussels, and amphibians of Tennessee in order to understand how human activities have spatially homogenized this biologically unique region.

2. THE ECOREGIONS OF TENNESSEE AND THEIR INFLUENCE ON FRESHWATER FAUNA

Tennessee boasts what is perhaps the most diverse assemblage of aquatic organisms in North America (Etnier and Starnes 1993). Contributing to its diversity is the juxtaposition of major ecoregions that transect the state (Griffith et al. 1997). Ecoregions are classified hierarchically according to spatial patterns and composition among biotic and abiotic factors including geology, physiography, hydrology, land use, soils, vegetation, and wildlife. Ecoregions are identified at four levels of resolution corresponding to Roman numerals I-V. Beginning at the coarsest level (i.e., Level I), North America contains 15 major ecoregions, and at a slightly higher resolution (Level II), North America is divided into 51 ecoregions. Griffith et al. (1997) subdivides Tennessee into seven Level III and 25 Level IV ecoregions. In this study, we used Level III resolution since it was broad enough to allow mapping of species ranges from existing atlases, yet specific enough to depict a gradient of landscape types across Tennessee. From west to east, the seven Level III ecoregions of the state include the Coastal Plains (CP), the Western Highland Rim (WHR), the Nashville Basin (NB), the Eastern Highland Rim (EHR), the Cumberland Plateau (CuP), the Ridge and Valley (RV), and the Blue Ridge Mountains (BR) (Figure 1). It should be noted that some authors (e.g., Etnier and Starnes 1993) have characterized Tennessee's landscape using physiography. Physiographic provinces contained within Tennessee are very similar to the ecoregions of Griffith et al. (1997); however, the Eastern and Western Highland Rim regions are combined to make one physiographic region called the Highland Rim (HR) which encircles the Nashville Basin. In this analysis, we retained Etnier and Starnes use of physiography in determining distribution of fishes; however, we used ecoregions based on Griffith et al. (1997) in assessing the ranges of

12. Spatial Homogenization of the Aquatic Fauna of Tennessee

mussels and amphibians. Regardless of the classification scheme, Tennessee's geographic situation, that of extending across multiple and distinct environmental regions, has not only produced a diverse aquatic fauna but also provides an intriguing setting for analyzing the spatial dynamics of biodiversity.

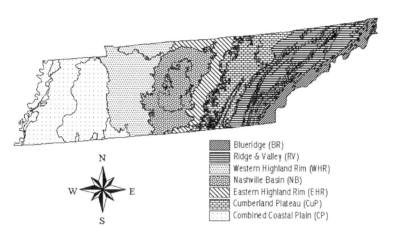

Figure 1. Ecoregions of Tennessee as presented by Griffith et al. 1997.

Examining major taxa such as fish, mussels, and amphibians illustrates the richly diverse aquatic ecosystems of Tennessee. Of fish alone, Tennessee's waters harbor 29 families within 18 orders, accounting for 293 species, most within the Percidae (darters and perches, 93 spp.) and Cyprinidae (minnows, 83 spp.)(Etnier and Starnes 1993). Parmalee and Bogan (1998) report approximately 150 species of freshwater mussels, 17 of which are believed to be extinct. Among amphibians, Tennessee contains 66 species in seven families within two orders. Over the past 65 years, the actions of such federal agencies as the Tennessee Valley Authority and United States Army Corps of Engineers have resulted in impounding virtually all of the medium to high order rivers in state (Etnier and Starnes 1993). The effects of these impoundments, exacerbated by silt from development, nutrient rich run-off from farming, acid mine drainage, and stresses of impervious surfaces due to urban sprawl have led to the federal or state listing of 51 fish, 39 mussels, and 12 amphibians (TDEC 1999).

In this analysis, we quantify the number of aquatic species within each ecoregion based on information published in Etnier and Starnes (1993) for fish, Parmalee and Bogan (1998) for mussels, and Redmond and Scott (1996) for amphibians. With the exception of fish, information on conservation status for each taxon was obtained from the Tennessee Department of Environment and Conservation (TDEC 1999). Conservation status for fish was extracted from Etnier and Starnes (1993). For a more detailed account of ecoregion descriptions, refer to Griffith et al. (1997).

3. QUANTIFYING SPATIAL DIVERSITY

Quantifying spatial homogenization of aquatic species across the ecoregions of Tennessee requires calculating within–ecoregion and between–ecoregion diversity. Within–ecoregion diversity is measured by simply tallying the number of species whose ranges are contained within a particular ecoregion. Between–ecoregion diversity is much more complex to calculate. Several authors have used slightly different formulas to answer various questions (e.g., Whittaker 1960, Cody 1986, MacArthur 1965) resulting in a hodge-podge of formulas (all typically called 'beta' diversity) each saddled with their own assumptions (Shmida and Wilson 1985, Huston 1994). We followed the method of Russell (1999) by defining between–ecoregion changes in diversity simply as spatial turnover. Spatial turnover can be calculated in exactly the same way as temporal turnover and is useful for comparing any kind of community difference in species composition (Russell 1999). Spatial turnover is measured by tallying the number of species "lost" or "gained" as one moves from one location to another along a spatial continuum. The resultant value is then standardized to the size of the community. We have slightly modified Russell (1999) to obtain the following formula for spatial turnover,

$$T_S = (G_S + L_S)/\alpha \qquad \text{(EQUATION 3.1)}$$

where S is a given ecoregional comparison (i.e., the spatial extent under consideration), G_S is the number of species found in the first ecoregion but not the second, L_S is the number of species found in the second ecoregion but not the first, and α is the total number of species found within both. Thus, T_S reflects the degree of faunal similarity between two adjacent ecoregions. Russell (1999) calculates α separately for each community 'state'. In this case these states are the two ecoregions under comparison. Russell (1999) then adds these two richness values together to arrive at the denominator. Instead we sum the total number of species in both 'states' and use this as the denominator. Thus, our method inevitably produces a lower turnover value than that obtained by the origional equation in Russell (1999) as we do not count species that are present in both ecoregions twice.

Spatial turnover will be high if the two ecoregions have evolved largely separate faunas due to ecological or physiographic isolation (e.g., drastically different habitat types, watershed divides). These pressures increase as the distance between ecoregions increases, resulting in greater and greater turnover values. We limit our ecoregion comparisons to only those adjacent to one another, thus minimizing this effect. Spatial turnover will be low if the ecoregions have many species in common. This latter scenario can be due to several natural reasons including historic stream capture, waterway connections, regular catastrophic flooding, climatic fluctuations, or merely by accident. It can also result from intentional or accidental widespread release of exotic species, or anthropogenic waterway connections (e.g., the Tenn-Tom Waterway, the Saint Laurence Seaway, etc.).

Following Harrison (1993) we calculated within– and between–ecoregion diversity at two different temporal states in order to understand how extinction and invasion are shaping regional faunas. The first temporal state is a historical scenario where all species that are believed to be extinct or locally extirpated are still extant and introduced species do not exist. This is intended to represent what the faunal assemblage may have been prior to significant human disturbance. The second temporal state is an estimated future scenario where extinct or locally extirpated species as well as those listed by federal or state agencies as threatened, endangered, or vulnerable are omitted from the data set. We make the assumption that future conservation measures, if taken for a given species, will be ineffective in saving the species from future decline. That is not to say that we believe all species currently deemed imperiled to some degree will ultimately become extinct. Likely, some will continue to survive; however, their survival as rare species will contribute negligibly to ecological and evolutionary processes in the future. Although numerous species-level and some ecosystem-level conservation efforts are underway (and some have been successful), the pressures of an ever-increasing human population combined with detrimental land use practices, make the long term prognosis for Tennessee's aquatic fauna appear bleak. Further, because exotic species can sometimes be controlled but rarely if ever eliminated from their novel environs, we include all currently established exotic species within the future scenario. We consider this a conservative approach to predicting the future of biological invasions. The rate of invasion for aquatic species is likely to increase in the future (Etnier and Starnes 1993) within Tennessee waters; however, it is difficult to predict which species will invade and which ecoregions will be invaded.

4. RESULTS

We begin by comparing historic and future statewide and within–ecoregion diversity values. A summary for the changes in within–ecoregion diversity for each taxon is provided in Table 1. Statewide diversity decreased for all taxonomic groups (12% for fish, 43% for mussels, and 18% for amphibians) when comparing historic and future scenarios. Currently, more than a quarter (25.7%) of the state's mussel species are vulnerable to extinction, as are nearly one-fifth of fish and amphibian species (17.4% and 18.2%, respectively). Twenty-two of the state's present-day 294 species of fish are considered exotic (7.5%) as are three of 136 mussels (2.2%). There are no known exotic amphibians established within the borders of Tennessee. There are believed to be at least two fish species that formerly inhabited the Tennessee drainage that are now extinct. The white-line topminnow (*Fundulus albolineatus*) is known only from one specimen collected at Big Spring near Huntsville, Alabama (Williams and Etnier 1982) on the lower bend of the Tennessee River. It is not included in the historic fauna of the state of Tennessee and was excluded from our analysis. Conversely, the harelip sucker (*Lagochila lacera*) is believed to have been widespread throughout several major drainages in eastern North America including the Tennessee and Cumberland until the late nineteenth century. No modern records of it are known and it is almost surely extinct. Etnier and Starnes (1993) provide thorough summary of its former distribution in Tennessee. While many species of amphibians are considered vulnerable to extinction, we know of none that have already vanished from the state's fauna. Unlike fishes and amphibians, freshwater mussels leave behind fossil evidence in the form of shells when they perish. As a result, it is known that at least 16 species (10.4%) no longer inhabit the waters of Tennessee. Further, it is possible, perhaps even likely, that other aquatic species may have disappeared from Tennessee's waters prior to being noticed by early naturalists.

Future within–ecoregion diversity decreased for nearly all taxa and ecoregions when compared to historic values. Exceptions included fish in Cumberland Plateau and Blue Ridge where within–ecoregion diversity increased by 1.9% and 7.3%, respectively, and mussels in the Coastal Plain where it increased by 5.7%. The Ridge and Valley ecoregion displayed the largest decline in within–ecoregion diversity among fish and mussels, dropping in total species richness from 160 to 148 for fish (representing an 8% loss), and from 106 to 45 (a 58% loss) in mussel species. This ecoregion has the highest number of vulnerable fishes (24) and mussels (31), but ranks among the lowest in the percentage of exotic species among these two groups (8% and 2%, respectively). Among amphibians, the largest decrease was in the Blue Ridge where species richness dropped from 50 to 42 species, a 16.0% decline.

Table 1. Historic to future projected changes in statewide and ecoregion diversity. See text for abbreviation definitions.

Amphibians

	CP	WHR	EHR	NB	CuP	RV	BR	Statewide
Past Species	34	38	38	33	41	38	50	66
Future Species	32	35	34	31	36	34	42	54
Percent change	-5.88	-7.89	-10.53	-6.06	-12.20	-10.53	-16.00	-18.18
Vulnerable Species	2	3	4	2	5	4	8	12
% vulnerable	5.88	7.89	10.53	6.06	12.20	10.53	16.00	18.18
Extinct	0	0	0	0	0	0	0	0
% extinct	0	0	0	0	0	0	0	0
Exotic	0	0	0	0	0	0	0	0
% exotic	0	0	0	0	0	0	0	0

Fish

	CP	HR	NB	CuP	RV	BR	Statewide
Past Species	128	164	92	52	160	64	272
Future Species	127	160	91	53	148	69	243
Percent Change	-0.79	-2.50	-1.10	1.89	-8.11	7.25	-11.93
Vulnerable species	13	16	5	5	24	6	51
% vulnerable	9.29	9.09	5.21	8.62	13.95	8.00	17.35
Extinct	0	1	1	0	1	0	1
% Extinct	0	0	0	0	0	0	0
Exotic Species	12	12	4	6	12	11	22
% Exotic	8.5	6.82	4.17	10.34	6.98	14.67	7.48

Mussels

	CP	WHR	EHR	NB	CuP	RV	BR	Statewide
Past Species	35	90	41	91	54	106	20	149
Future Species	37	58	34	48	30	45	13	85
Percent change	5.71	-35.56	-17.07	-47.25	-44.44	-57.55	35.00	-42.95
Vulnerable species	0	19	6	21	13	31	2	35
% vulnerable	0.00	22.35	14.63	25.93	26.53	33.70	11.11	25.74
Extinct	0	8	1	12	6	16	3	16
% Extinct	0.00	8.89	2.44	13.19	11.11	15.09	15.00	10.74
Exotic	2	3	1	2	1	2	1	3
% exotic	5.41	3.53	2.44	2.47	2.04	2.17	5.56	2.21

Between–ecoregion diversity, or spatial turnover, decreased for all taxa/ecoregion combinations excluding amphibians between the Coastal Plain and Western Highland Rim where it increased by 38.8% (Table 2). Statewide, average spatial turnover decreased for all groups, ranging from a 16% drop in fishes to a 20% drop in amphibians. Maximum declines in spatial turnover among amphibians and fish occurred between the Cumberland Plateau and the Ridge and Valley where amphibian turnover dropped 62.4%, and fish turnover decreased by 23.3%. The maximum decrease in spatial turnover among mussels occurred between the Cumberland Plateau and the Eastern Highland Rim with a 35.4% decrease. Historically, spatial turnover is relatively low for mussels and amphibians indicating a high level of natural similarity between ecoregions; however, a clear trend of increasing spatial turnover exists among historical fish faunas moving from west to east across the state. This large break in fish faunal similarity is particularly apparent as one moves from the Highland Rim ecoregion into the Cumberland Plateau. This rapid turnover in fish species continues as one moves eastward. Overall, if we assume that vulnerable taxa will ultimately become extinct and exotic species will continue to thrive, then it appears that the ecoregions of the western portion of the state will continue to support aquatic fauna that maintain a high degree of similarity. That said, it is equally true that the historically diverse faunas of central and eastern Tennessee will lose their distinctiveness.

5. DISCUSSION

Tennessee maintains a rich aquatic natural heritage, the face of which has been drastically altered largely as the result of human activities throughout the twentieth century. Mussels appear to be taking the brunt of the punishment. The loss of mussels from Tennessee and neighboring states stands among the most alarming extinction events in modern history (Williams et al. 1992). This loss of mussels, and other aquatic organisms, is primarily the result of the construction of numerous dams, primarily from 1935 on, that rapidly converted dynamic and complex flowing systems to a series of sluggish reservoirs lacking in habitat diversity. Today, most of the state's medium-sized rivers are entombed by reservoirs (Etnier 1997), leaving their habitats and faunas extremely vulnerable to extinction. In addition, poor (and largely unregulated) agricultural and silviculture practices, in tandem with the effect of urban sprawl, continue to destroy isolated spring habitats and headwater creeks (Etnier 1997).

Unlike traditional assessments of biodiversity, biotic homogenization is an attempt to better understand the entire effect on a given fauna by including both species decline and invasions. By including exotic species in our analysis of richness declines, fish diversity remains comparatively high with respect to amphibians and (especially) mussels. As Marchetti et al. (this volume) point out, the inclusion of exotics in standard biodiversity calculations complicates our view of how to interpret such measures. For example, Marchetti et al. conclude that the introduction of novel species appears to be a driving force behind biotic homogenization among California's fish fauna. This provides and intriguing contrast to our results in that there are comparatively few exotics among Tennessee's aquatic fauna, an undoubted artifact of being a landlocked state. The aquatic fauna of Tennessee still decreases in biodiversity when we include exotic species in our counts; however, it tends to mask the loss of regional distinctiveness (see below).

Table 2. Historic to future projected changes in spatial turnover within Tennessee's ecoregions. See text for abbreviation definitions.

Amphibians

	Historic	Future	Deviation from Historic	Percent Deviation from Historic
CP-WHR	0.20	0.28	0.08	38.89
WHR-NB	0.31	0.24	-0.07	-21.41
NB-EHR	0.27	0.22	-0.05	-17.17
EHR-CuP	0.20	0.14	-0.07	-33.93
CuP-RV	0.41	0.15	-0.26	-62.39
RV-BR	0.37	0.31	-0.06	-16.00
Mean	0.31	0.21	-0.10	-30.18

Fish

	Historic	Future	Deviation from Historic	Percent Deviation from Historic
CP-HR	0.67	0.65	-0.03	-3.75
HR-NB	0.97	0.84	-0.13	-13.25
HR-CuP	2.89	2.24	-0.64	-22.22
CuP-RV	2.69	2.06	-0.63	-23.34
RV-BR	2.19	1.72	-0.47	-21.45
Mean	1.88	1.50	-0.38	-16.80

Mussels

	Historic	Future	Deviation for Historic	Percent Deviation from Historic
CP-WHR	0.71	0.56	-0.15	-21.57
WHR-NB	0.15	0.13	-0.02	-12.65
NB-EHR	0.57	0.42	-0.16	-27.00
EHR-CuP	0.52	0.33	-0.18	-35.35
CuP-RV	0.53	0.43	-0.10	-19.46
RV-BR	0.83	0.73	-0.10	-11.74
Mean	0.55	0.43	-0.12	-21.29

On a non-political spatial scale (i.e., ecoregions), future losses in species diversity are non-random in that they are not spread evenly through space. Within the Cumberland Plateau and Blue Ridge ecoregions future fish diversity actually increased. In this ecoregion, lost fishes are 'replaced' by exotic fishes. However, this is not a niche-for-niche ecological replacement. That is, most fish that are lost inhabit medium-sized rivers or small springs whereas most exotics live in lentic systems (Etnier 1997), making the long-term effect of homogenization a spatially non-random event. In short, we are witnessing a transition from flowing water habitats to standing water habitats (i.e., lotic systems are replaced by lentic systems). This change in habitat is, to a degree, mirrored by a change from lotic-adapted to lentic-adapted species and a wholesale replacement of functional groups, community structure, and ecological dynamics.

Similarly, future mussel diversity increased in the Coastal Plain. This can be attributed in part to the fact that mussel diversity is generally low in this ecoregion relative to the remainder of the state and that no species are listed as vulnerable to extinction. Hence, the primary long-term impact on this fauna will likely be due to exotic species. Of course, because the fauna is historically few in number, inter-specific interactions such as interference competition with zebra mussels, may induce extinction particularly for those species which may be inherently rare.

Most future losses in diversity occur within the Ridge and Valley ecoregion for mussels and fish and within the Blue Ridge for amphibians. We discuss the possible causes of sharp declines in diversity within the Ridge and Valley below. Among amphibians, future drops in within–ecoregion diversity generally increased in magnitude from west to east. The reason for this trend is unclear but may be related to habitat complexity. Western ecoregions are characterized by large, sluggishly flowing systems (similar to reservoirs). In the east, habitats are more variable and include numerous seeps, springs, and small creeks. As this habitat variety is diminished, species that are co-evolved for a thin range of habitat types are lost first.

Finally, future spatial turnover decreases in almost all ecoregion comparisons and across all taxa. Maximum decreases in amphibian and fish spatial turnover occurred between the Cumberland Plateau and Ridge and Valley ecoregions. Most of this decrease may be attributed to the large projected future declines in amphibian and fish diversity within the Ridge and Valley ecoregion. A very small portion of this ecoregion contains the unique Consasauga watershed. The headwater forests of this watershed remain largely untouched (Etnier and Starnes 1993), thus this region contains the last stronghold of several threatened fish and amphibians (and mussels). If we assume these unique species do not survive into the future, spatial turnover inevitably drops. This suggests that if we were to know the pre-impoundment fauna of the Mississippi and Tennessee watersheds (i.e. the primary river systems that make up the remaining ecoregions) the future decrease in spatial turnover within other ecoregion comparisons would have likely been greater.

For mussels, maximum decreases in spatial turnover occur between the Eastern Highland Rim and the Cumberland Plateau. The Cumberland Plateau region suffers from a recent and continued history of extreme land use change, most notably the presence of coal mining. A particularly devastating result has been the strip-mining of many of the region's watersheds over the last 30 years. In some cases, strip-mining has resulted in the virtual elimination of a watershed's aquatic fauna (Etnier and Starnes 1993). Today, acid mine drainage is a limiting factor in addition to continued strip-mining. The loss of several unique species within this ecoregion, combined with similarly high losses in the Eastern Highland Rim, essentially leaves a depauperate mussel fauna of broadly adapted, wide ranging species that can successfully survive in highly altered waterways (Parmalee and Bogan 1998).

The presence of several spatially common exotic fish and mussels further drives future turnover values down. With an increase in commercial river traffic, intentional sport fish stocking, and the dumping of bait buckets several exotic fish and mussels have colonized almost all of Tennessee's ecoregions. This is very apparent among the fishes. Sport fishes such as the walleye (*Stizostedion vitreum*) and rainbow trout (*Oncorhynchus mykiss*) are intentionally stocked throughout Tennessee waters. An entire suite of minnows (family Cyprinidae) have been transported as bait fish, and the ever-ubiquitous goldfish (*Carassius auratus*) has become well-established probably as an intentionally released aquarium fish. These exotic fishes decrease the spatial diversity of Tennessee's waterways simply by existing throughout the state. However, they may further decrease spatial diversity by directly causing the decline of many of their native congeners either by predation or competition (Etnier and Starnes 1993).

The main impact of aquatic habitat alteration and land use change on Tennessee waters has been the loss of regional distinction and an overall decrease in species richness. To appreciate such an effect may significantly aid our efforts to offset the current extinction crisis; however, to do so takes a broader view than that of traditional population-based conservation biology. By adopting a landscape– or watershed–based perspective toward land management, planning, and species conservation, we may be able to stabilize and avert future the damage. Measuring diversity between–ecoregions (i.e., calculating spatial turnover) can not only determine whether regional faunas are increasing or decreasing in similarity, it can also identify conservation "hotspots" whereby efforts can be targeted toward specific local regions.

ACKNOWLEDGEMENTS

We sincerely thank Drs. D.A. Etnier, M.P. Marchetti, M.L. McKinney, and G.J. Russell for their insightful reviews and comments. J.R. Duncan was supported as a Graduate Research Assistant by the University of Tennessee Water Resources Research Center.

REFERENCES

Allen, J.D. and A.S. Flecker. 1993. Biodiversity conservation in running waters: identifying the major factors that affect destruction of riverine species and ecosystems. Bioscience 43:32-43.

Angermeier, P.L. 1995. Ecological attributes of extinction-prone species: loss of freshwater fishes of Virginia. Conservation Biology 9(1):143-158.

Carlton, J.T. 1992. Dispersal of living organisms into aquatic ecosystems as mediated by aquaculture and fisheries activities. In: Dispersal of Living Organisms into Aquatic Habitats (A. Rosenfield and R. Mann, eds.), Maryland Sea Grant Publication, College Park, Maryland, Pp: 13-46.

Carlton, J.T. 1996. Biological invasions and cryptogenic species. Ecology 77(6):1653-1655.

Cody, M.L. 1986. Diversity, rarity, and conservation in Mediterranean-climate regions. In: Conservation Biology: the science of scarcity and diversity (M.E. Soule, ed.). Sinauer, Sunderland, MA.

Etnier, D.A. 1997. Jeopardized southeastern freshwater fishes: A search for causes. In: Aquatic Fauna in Peril (Benz, G.W. and D.E Collins, eds.), Southeast Aquatic Research Institute, Chattanooga, Tennessee.

Etnier, D.A. and W.C. Starnes. 1993. The Fishes of Tennessee. University of Tennessee Press. Knoxville, Tennessee.

Griffith, G.E., Omernik, J.M., and S.H. Azevedo. 1997. Ecoregions of Tennessee. EPA/600/R-97/022. Western Ecology Division, National Health and Environmental effects Research Laboratory, U.S. Environmental Protection Agency, Corvalis, Oregon.

Harrison, S. 1993. Species diversity, spatial scale, and global change. In: Biotic interactions and global change (P. Karieva, J.G. Kingsolver, and R.B. Huey, eds). Sinauer Associates, Inc. Sunderland, MA.

Huston, M.A. 1994. Biological diversity : the coexistence of species on changing landscapes. Cambridge University Press, New York.

Lockwood, J.L. 1999. Using taxonomy to predict success among introduced avifauna: relative importance of transport and establishment. Conservation Biology 13(3):560-567.

Lockwood, J.L., T. L. Brooks and M.L. McKinney. 2000. Taxonomic homogenization of the global avifauna. Animal Conservation 3:27-35.

Lodge, D.M. 1993. Biological Invasions: Lessons for Ecology. Trends in Ecology and Evolution 8:133-136.

MacArthur, R.H. 1965. Patterns of species diversity. Biological Review 40:510-533.

McKinney, M.L. 1998. On predicting biotic homogenization: species-area patterns in marine biota. Global Ecology and Biogeography Letters 7:297-301.

McKinney, M. and J.L. Lockwood 1999. Taxonomic patterns in the next mass extinction. Trends in Ecology and Evolution 11:450-453.

Parmalee, P.W. and A.E. Bogan. 1998. The Freshwater Mussels of Tennessee. University of Tennessee Press. Knoxville, Tennessee.

Redmond, W.H. and A.F. Scott. 1996. Atlas of Amphibians in Tennessee. Miscellaneous Publication Number 12, The Center for Field Biology, Austin Peay State Univesity. Clarksville, Tennessee.

Russell, G.J. 1999. Turnover dynamics across ecological and geological scales. In: Biodiversity dynamics: turnover of populations, taxa, and communities (McKinney, M.L. and J.A. Drake, eds.), Columbia University Press, New York.

Shmida, A. and M.V. Wilson. 1985. Biological determinants of species diversity. Journal of Biogeography 12:1-20.

Tennessee Department of Environment and Conservation: Division of Natural Heritage website http://www.state.tn.us/environment/nh/

Whittaker, R.H. 1960. Vegetation of the Siskiyou Mountains, Oregon and California. Ecological Monographs. 30:279-338.

Williams, J.D. and D.A. Etnier. 1982. Description of a new species, *Fundulus julisia*, with a redescription of *Fundulus albolineatus* and a diagnosis of the subgenus Xenisma (Teleostei: Cyprinodontidae). Occasional Papers of the Museum of Natural History, The University of Kansas, Lawrence, Kansas.

Williams, J.D., M.L. Warren Jr., K.S. Cummings, J.L. Harris, and R.J. Neves. 1992. Conservation status of the freshwater mussels of the United States and Canada. Fisheries 18(9):6-22.

Chapter 13

Homogenization of California's Fish Fauna Through Abiotic Change

Michael P. Marchetti[1], Theo Light[1], Joaquin Feliciano[1], Trip Armstrong[1], Zeb Hogan[1], Joshua Viers[2], and Peter B. Moyle[1]

[1]*Department of Wildlife, Fish and Conservation Biology, University of California, Davis CA, 95616, USA*
[2]*Information Center for the Environment, University of California, Davis, CA, 95616, USA*

1. INTRODUCTION

The decline of native fish populations and the invasion of non-native fishes are the most noticeable trends in California's freshwater fish assemblages over the last century (Moyle and Williams 1990, Moyle 2000). Moyle (2000) and Dill and Cordone (1997) date the first introduction of non-native fish into California back to the latter half of the 19th Century. Yoshiyama et al. (1998) place the beginning of the decline of the state's chinook salmon populations also near the turn of the century. The mid-1800's also marks the beginning of a population explosion in California, driven by the discovery of gold in the Sierra Nevada in 1848.

As a result of development and settlement forces following this discovery, California's aquatic habitats have been substantially altered. Based on our observations of California's freshwater fish fauna, we believe that changes in the composition of fish assemblages and the alteration of aquatic habitats are strongly linked. We postulate that a physical convergence of aquatic habitats (abiotic homogenization), fueled by state-wide forces of development and anthropogenic change, facilitates increases in faunal similarity (biotic homogenization).

By combining state-wide data on historic and present fish assemblages with data on land use and development patterns, we investigated potential links between abiotic homogenization and biotic homogenization within California. We examined changes to California's fish fauna at a number of different scales: whole zoogeographic provinces, watersheds within the same zoogeographic province and, at the finest scale, individual stream reaches after reservoir construction. In addition, we used multiple regression to examine the relationship between physical watershed characteristics and changes in faunal assemblages within California's watersheds.

2. BIOLOGICAL SETTING

The majority of California's land area is bounded on the north, south and east by either large mountains or large rivers, and on the west by the Pacific Ocean. These boundaries have created natural borders to the state as well as producing a high diversity of terrain and ecologically distinct habitats. The composition and distribution of California's original fish fauna reflect the state's geographic and hydrologic diversity (Moyle 2000). Although California has a relatively small number of native freshwater fish species (66), it has a disproportionately large number of endemic taxa (24), due in part to the geographic isolation of the state (Moyle 2000). Furthermore, the native fish assemblages are often regionally quite distinct. For instance, of the 15 fish taxa originally present in Southern California, 6 are endemic only to that region (Moyle 2000).

For a variety of reasons, California has been invaded by a large number of non-native fish species (Dill and Cordone 1997, Moyle 2000). There are currently 110 recorded freshwater fish species in California, a 76% increase from the original 66 species (Moyle 2000). This total number of species represents 50 successful non-native species introductions and 6 native species extinctions (Moyle 2000). Ironically, the species-poor and distinctive nature of California's fish fauna makes the documentation and tracking of fish introductions and extinctions relatively easy even across watersheds. Dill and Cordone (1997) were able to assemble an introduction history for every exotic fish found in the state.

Within California's boundaries, there are several aquatic zoogeographic provinces (Moyle and Cech 1996, Moyle 2000), each with distinctive climates, tectonic settings, and hydrologic regimes (Mount 1995). We have chosen to focus on six provinces defined by Moyle (2000): Klamath, Sacramento-San Joaquin, North Coast, Great Basin, South Coast, and the Salton Sea Basin (the only included watershed of the Colorado River Province) (Figure 1). Moyle's divisions, based primarily on freshwater fish distributions, are comparable to divisions based solely on hydrologic factors (Mount 1995, Moyle 2000).

13. Homogenization of California's Fish Fauna

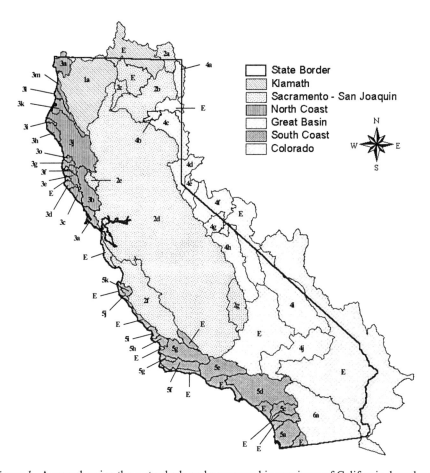

Figure 1. A map showing the watersheds and zoogegraphic provinces of California, based on delineations in Moyle (2000). (1) Klamath zoogeographic province: a) Lower Klamath River; (2) Sacramento – San Joaquin zoogeographic province: a) Goose Lake, b) Pit River, c) McCloud River, d) Sacramento–San Joaquin River, e) Clear Lake, f) Monterey Basin, g) Kern River; (3) North Coast zoogeographic province: a) Tomales Bay, b) Russian River, c) Gualala River, d) Garcia River, e) Navarro River, f) Big River, g) Noyo River, h) Matolle River, i) Bear River, j) Eel River, k) Mad River, l) Little River, m) Redwood Creek, n) Smith River, o) Ten Mile Creek; (4) Great Basin zoogeographic province: a) Surprise Valley, b) Eagle Lake, c) Susan River, d) Truckee River, e) Carson River, f) Walker River, g) Mono Lake, h) Owens River, i) Amargosa River, j) Mojave River; (5) South Coast zoogeographic province: a) San Diego River, c) Santa Margarita River, d) Los Angeles River, e) Santa Clara River, f) Santa Inez River, g) Santa Maria River, h) San Luis Obispo Creek, i) Morro Bay, j) Big Sur Coast, k) Carmel River; (6) Colorado zoogeographic province: a) Salton Sea Basin. 'E' indicates watersheds that were excluded from the analyses.

3. CULTURAL SETTING

Widespread urbanization and development have substantially reduced the hydrologic diversity among the state's zoogeographic provinces. Some of the most dramatic alterations have been to the state's waterways themselves (i.e., dams, diversions, and channelization) (Reisner 1986, Mount 1995). Since the mid-1800's, Californians have erected over 1400 dams with a storage capacity of approximately 42 million acre feet, utilizing close to 60% of the state's average annual runoff (Mount 1995). These dams change the quantity and variability of flow in downstream reaches and create lake environments in regions that were previously free-flowing (Petts 1984, Power et al. 1996). In addition, vast amounts of water are transferred around the state by a complex water system comprising over 140 aqueducts and canals. The California Aqueduct alone carries over 2.5 million acre-feet annually from the Sacramento-San Joaquin Delta hundreds of miles south to the southern regions of the state (Mount 1995). Stream channels that are not dammed or diverted are often straightened or paved in attempts to reduce flooding, reduce groundwater loss and increase flood drainage in urban areas (Lazaro 1979, Mount 1995).

Streams and natural waterways can also be affected in an indirect manner through general development in the watershed (urban growth, logging, agriculture, etc.) (Richards et al. 1996). Changes to waterways as a result of urbanization and development are well documented and include, but are not limited to: increases in the duration and total flow of storm runoff events (Espey 1969, Lazaro 1979, Mount 1995), increases in sedimentation and decreases in shade and large woody debris following timber harvesting (Chamberlin et al. 1991, Holtby 1988), increases in sedimentation from road construction (Furniss et al. 1991), increases in eutrophication and concentration of pesticides and heavy metals in waterways from agriculture (Mount 1995), and increases in bank erosion, eutrophication, and sedimentation from grazing pressure (Platts 1991).

There is also evidence demonstrating that the changes to aquatic habitats affect resident fish assemblages (Scott et al. 1986, Weaver and Garman 1994, Poff and Allan 1995, Knapp and Matthews 1996, Bennett and Moyle 1996, Onorato et al. 1998). Anthropogenic changes can cause abiotic homogenization in two distinct ways. First, these changes increase the physical homogeneity within a specific stream reach by reducing habitat complexity. Second, similar patterns of watershed development around the state reduce heterogeneity among the state's zoogeographic provinces. These alterations may promote biotic homogenization by making waterways more hospitable for introduced fishes and less hospitable for some native species.

4. METHODS

4.1 Fish Distribution and Species Richness

Fish distribution by watershed data were compiled using Moyle (2000), which contains extensive zoogeographic and watershed specific species lists. From these data, we calculated species richness (number of species) for each watershed, both pre-1850 and in its current form, as well as the change in the number of species between these two time periods (Table 1). Some watersheds within the state were excluded from the analysis either because there are no fish in the watershed, or because the watershed extends significantly outside the state boundaries (Figure 1).

To compile the presence/absence data on current and pre-reservoir fish assemblages, we relied on information from private and state fish biologists (T.L. Ford, R.E.Leidy, J. Setka, L. Pardy, R.A. Swenson, F. Kopperdahl pers. com.). For the current fish assemblages in reservoirs, we used species lists compiled from multiple years of sampling that were not obviously biased towards reporting only game fish (i.e., families Centrarchidae and Salmonidae). We recorded only presence or absence of fish species, and we included only those species with reproducing populations, ignoring species present through continual stocking. In one unique case, we included striped bass (*Morone saxatilis*) in Silverwood Lake due to its large annual presence, despite the fact that the species is maintained by continual introduction through the California Aqueduct from the San Francisco Bay Delta. To generate the species lists for rivers prior to damming, we relied on the expert opinions of two regional fish biologists (P.B. Moyle, R.E. Leidy) and Moyle (2000).

4.2 Similarity comparisons

We examined homogenization of California's fish faunas by examining the pairwise similarity among faunas on three scales: between hydrologic provinces, between watersheds within hydrologic provinces, and between individual reservoirs. For each pair of provinces/watersheds/reservoirs we calculated the Jaccard Index of Similarity between their modern fish faunas (henceforth referred to as Jaccard faunal similarity score - JFS score).

The Jaccard Index is independent of sample size and the number of species in each assemblage and has values ranging from 0 (when no species are found to be in common) to 1 (when all species are shared by both assemblages). The Jaccard formula is $S_j = a/(a+b+c)$ where:

Table 1. Species richness for each California watershed. Data were compiled from Moyle (2000).

Watershed	Code	Original Native	Extinct Native	Intro.	Current	Change
Lower Klamath R.	KLA.1a	20	1	14	33	13
Goose Lake	GOS.2a	8	0	11	19	11
Pit River	PIT.2b	13	1	15	27	14
McCloud River	MCL.2c	7	4	4	7	0
Sacramento/ San Joaquin River	SAC.2d	29	4	41	66	37
Clear Lake	CLE.2e	14	3	18	29	15
Monterey Basin	SAL.2f	19	3	20	36	17
Kern River	KER.2g	4	0	7	11	7
Tomales Bay	TOM.3a	11	0	7	18	7
Russian River	RUS.3b	21	1	19	39	18
Gualala River	GUA.3c	8	0	0	8	0
Garcia River	GAR.3d	8	1	0	7	-1
Navarro River	NAV.3e	9	0	0	9	0
Big River	BIG.3f	8	0	0	8	0
Noyo River	NOY.3g	5	0	0	5	0
Matolle River	MAT.3h	8	0	0	8	0
Bear River	BEA.3i	9	0	0	9	0
Eel River	EEL.3j	14	2	10	22	8
Mad River	MAD.3k	14	2	8	20	6
Little River	LIT.3l	9	0	0	9	0
Redwood Creek	RED.3m	12	0	6	18	6
Smith River	SMT.3n	12	0	0	12	0
Ten Mile Creek	TEN.3o	7	0	0	7	0
Surprise Valley	SUP.4a	3	0	2	5	2
Eagle Lake	EAG.4b	5	0	2	7	2
Susan River	SUS.4c	8	1	7	14	6
Truckee River	TRU.4d	8	0	15	23	15
Carson River	CAR.4e	8	0	14	22	14
Walker River	WAL.4f	8	0	13	21	13
Mono Lake	MON.4g	0	0	6	6	6
Owens River	OWE.4h	4	0	14	18	14
Amargosa River	AMA.4i	3	0	2	5	2
Mojave River	MOJ.4j	1	0	23	24	23
San Diego River	DIE.5a	7	1	26	32	25
Santa Margarita R.	MGT.5c	9	3	12	18	9
Los Angeles River	ANA.5d	12	2	34	44	32
Santa Clara River	CLA.5e	7	0	24	31	24
Santa Inez River	YNE.5f	6	0	16	22	16
Santa Maria River	MAR.5g	7	0	8	15	8
San Luis Obispo Creek	SLO.5h	7	0	8	15	8
Morro Bay	MOR.5i	8	0	10	18	10
Big Sur Coast	SUR.5j	6	0	0	6	0
Carmel River	CML.5k	5	0	12	17	12
Salton Sea Basin	STN.6a	1	0	24	25	24

a = number of species in both samples,
b = number of species in sample A, but not sample B,
c = number of species in sample B, but not sample A.

In cases where both compared watersheds were originally fishless, we assigned them a JFS score of zero.

We repeated this calculation on the pre-1850 fish faunas (native fishes plus known extinct natives). To determine whether similarity among faunas has, in general, increased or decreased since 1850, we plotted current JFS scores against past JFS scores, noting how many points fell above and below the 1-1 line. Points above the unity line indicate that the two watersheds in question are more similar in the present then in the past (biotic convergence or homogenization), while points below the line indicate the opposite, that the two watersheds were more similar in the past then in the present (divergence) (Figure 2). We then used a sign test to compare the number of positive with the number of negative changes. However, multiple comparisons using the same watershed compromise statistical independence and increase N for the sign test to $N^2/2$, potentially resulting in an inflated significance value. We therefore repeated the sign test on the set of average JFS scores for each watershed/province/reservoir for a more conservative estimate of P. While this does not entirely eliminate non-independence of points (each shares one comparison in common with all other points), it does reduce N to equal the number of sites in the original data.

Reservoir data were limited to the two provinces for which we had data from multiple reservoirs (Southern California and the Central Valley). We also compared each reservoirs to all other reservoirs to examine larger-scale variation.

4.3 Regression analysis

To explore the relationship between physical homogenization of habitat and faunal change, we examined the effect of 11 environmental variables on the JFS scores between each watershed's original (pre-1850) fish fauna and its current fish fauna. The JFS score between past and present faunas is an index of faunal constancy (or lack of change) within each watershed, not, as above, of similarity among watersheds. Eleven environmental measures for each watershed (compiled using GIS, J. Viers unpublished data) were selected for regression analysis as relatively independent individual measures (Table 4). We used our best biological judgment, based on our knowledge of fish-environment relationships in California, to select representative variables. We transformed variables for normality (Table 4), and centered, but did not standardize, variables for ease of interpretation.

We first calculated the Pearson's correlation coefficients and associated Bonferroni-adjusted probabilities to assess univariate relationships between each independent variable and the JFS score of past to present faunas. We then did a complete-model multiple regression, including all 11 independent variables. We did not include interaction terms in the complete model because the number of variables necessary would have exceeded the number of cases (watersheds) available. Finally, we used forward and backward stepwise selection to estimate a reduced model. We then tested all first-order interactions among terms retained in the reduced model. Significant interactions were retained in the final model.

5. RESULTS

5.1 Species Richness

Looking within zoogeographic provinces, most watersheds experienced net gains in species richness (Table 1). Within the Sacramento/San Joaquin province six of seven watersheds had net gains in species and one had no change (although this included four extinctions and four introductions). Within the North Coast province, five out of the 15 watersheds had net gains, nine had no change and one had a loss of one species. The Great Basin province had net gains in species in all 10 watersheds, and the South Coast province had net gains in nine of its 10 watersheds while one showed no change.

Table 2. Species richness values for each zoogeographic province. Data were compiled from Moyle (2000).

Province	Code	Original Native	Extant Native	Current	Intro.	Extinct	Incr.	% Incr.
Klamath	1a	20	19	33	14	1	13	165
Sacramento/ San Joaquin	2a-g	36	33	73	40	3	37	203
North Coast	3a-o	26	24	50	26	2	24	192
Great Basin	4a-j	13	12	42	30	0	29	323
South Coast	5a-k	15	14	52	38	1	37	347
Salton Sea	6a	1	1	25	24	0	24	2500

When a similar tally was made at the level of the zoogeographic province (Table 2), all provinces experienced a net gain in species. The largest number of species gained was shared by the Sacramento/San Joaquin province and the South Coast province, both with a net gain of 37 species. The largest percentage change was in the Salton Sea, which exhibited an incredible 2500% increase in fish species in the last two centuries.

5.2 Similarity comparisons

Comparisons of JSF scores between past and present fish faunas yielded strikingly different results depending on the scale (zoogeographic provinces vs. watersheds) and type (watersheds vs. reservoirs) of comparison. At the largest scale, faunal similarity increased among all zoogeographic providence comparisons since 1850 (Figure 2, Table 3). This indicates a highly significant increase in homogeneity of fish faunas among zoogeographic provinces (sign test, 2-sided $P<.0001$). In the more conservative sign test, all six average values also increased; in this case only the one-sided sign test remains significant at $P<.05$.

Table 3. Number of increases (positives) and decreases (negatives) in JFS scores for watershed and reservoir comparisons, between 1850 and the present.

	# of Positives	# of Negatives	# of no change	two sided P value
Watershed Comparisons				
-within zoogeographic provinces				
Sacramento/San Joaquin	14	7	0	0.189
Great Basin	22	22	1	1.0
South Coast	3	52	0	<0.0001
North Coast	21	70	14	<0.0001
-between zoogeographic provinces				
All	15	0	0	<0.0001
Reservoir Comparisons				
-within zoogeographic provinces				
Sacramento/San Joaquin	66	9	2	<0.0001
South Coast	14	14	0	1.0
-between zoogeographic provinces				
All	4	2	0	0.687
-all reservoirs				
All	231	51	17	<0.0001

At the scale of watersheds within provinces, current similarity either decreased or showed no clear pattern. In both the North Coast and South Coast provinces, similarities among watersheds were higher in the past (Figure 3, Table 3, 2-sided $P<.0001$ for each). In the more conservative sign test, all 15 watersheds in the North Coast and all 11 watersheds in the South Coast showed decreased average similarity values, again resulting in significant sign tests (2-sided $P<.0001$ and $P<.001$, respectively). In both the Sacramento/San Joaquin and Great Basin provinces, comparisons were mixed, with 14 positive and seven negative changes in the Sacramento/San Joaquin province, and 22 in each direction (and one with no change) in the Great Basin. None of the sign tests were significant for these two provinces.

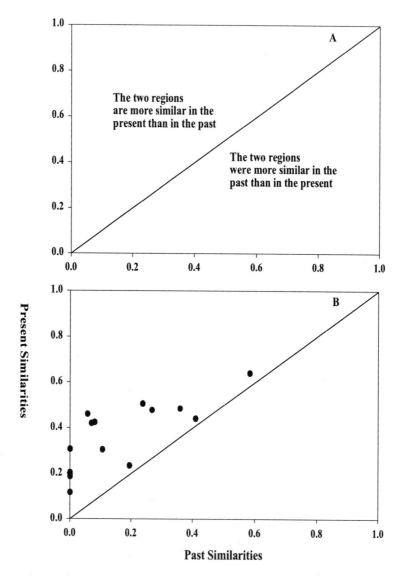

Figure 2. The top graph is a general interpretation scheme for the graphs utilizing Jaccard faunal similarity scores (JFS scores). The bottom graph is a pairwise comparison of present and past similarities of zoogeographic provinces in California using JFS scores.

13. Homogenization of California's Fish Fauna

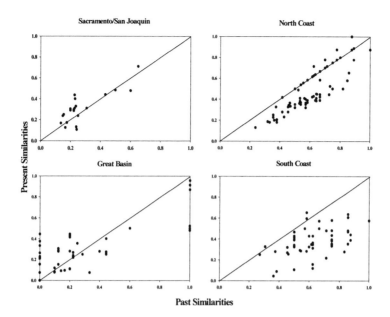

Figure 3. Pairwise comparisons of present and past similarities of watersheds within each zoogeographic province using Jaccard faunal similarity scores (JFS scores).

The Klamath and Salton Sea provinces could not be examined as they contained only one watershed each.

We had data to make within-province reservoir comparisons for only two zoogeographic provinces, Sacramento/San Joaquin and South Coast. Reservoirs in the Sacramento/San Joaquin province are more similar in the present than the streams and river reaches they replaced (Figure 4, Table 3; P<.001 on both tests), while reservoirs in the South Coast province showed an equal number of present and past similarities (Figure 4, Table 3). The comparison of all reservoirs to all other reservoirs also showed increased overall present similarity (Figure 4, Table 3; 2-sided P<.0001).

5.3 Regression analysis

Seven of the 11 environmental variables were significantly correlated (Bonferroni adjusted $P < .05$) with the JFS scores between past and present faunas (Table 5). All four water development variables, as well as urban development, were negatively associated with the JFS scores, suggesting the more a watershed was influenced by these activities, the less its current fish fauna resembled that of 150 years ago. Proportion agriculture was marginally negatively associated with the JFS scores as well (P<.1). Watershed area and mean elevation were strongly negatively associated with the JFS scores (P < .001), while only mean rainfall showed a strong positive relationship with JFS scores (P < .001).

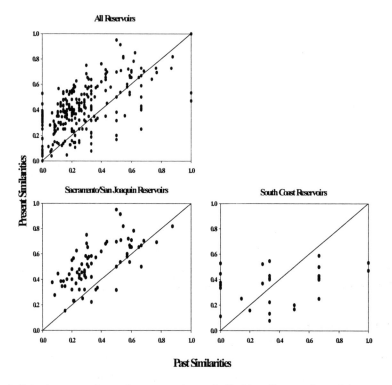

Figure 4. Pairwise comparison of present and past similarities of reservoirs and the streams they replaced within the Sacramento/San Joaquin, South Coast and all reservoirs using Jaccard faunal similarity scores (JFS scores).

The complete multiple regression model was highly significant, explaining 72% of the variance in JFS scores (F = 11.02, Adj. R^2 = 0.72; Table 5). In this model, the direction of the strongest effects remained consistent with the univariate relationships: aqueduct density and mean elevation had significantly negative effects on JFS scores, while mean rainfall remained the only significant positive effect. These three variables were retained in the best stepwise model, along with the variable dams (also with a negative coefficient). Both forward and backward selection produced the same model. Two first-order interactions proved significant: dams*aqueducts (P < .05) and rainfall*elevation (P < .01). These were included in the reduced model, which then had six variables and explained 82% of the variance in the JFS scores (F = 33.83, P<.0001; Table 2).

Stepwise multiple regression is an imperfect process, and it is important to acknowledge that other models could be built from these data that explain a similar percent of the variation in the similarity measure. For example, reservoir area can be substituted for dams in the above model with almost identical results (F=33.59, Adj. R^2=.8197, P<.0001), and proportion developed can be substituted for dams and aqueduct density (F=29.09, Adj. R^2=.72, P<.0001).

Table 4. Environmental variables used in correlation and multiple regression analysis. Variables were transformed as indicated when necessary to satisfy assumptions of normality. Reservoir area for the Truckee River watershed, a large outlier (possibly due to the inclusion of Lake Tahoe), was estimated from a regression of reservoir area on dams and dam volume for all other watersheds. High protection status includes national parks, wilderness areas and reserves.

Variable	Units	Transformation
Dams	#/1000 km^2	sqrt(x+1)
Reservoir area	100 m^2/km^2	arcsine-sqrt
Ditch density	m/km^2	log(x+1)
Aqueduct density	m/km^2	log(x+1)
Proportion developed	proportion	arcsine-sqrt
Proportion agriculture	proportion	arcsine-sqrt
High protection status	proportion	arcsine-sqrt
Mean annual rainfall	mm	log(x)
Mean elevation	m	log(x)
Stream density	km/km^2	none
Watershed area	m^2	log(x)

5.4 Discussion

Our evidence suggests that California's freshwater regional fish faunas have become more similar since 1850. This homogenization process (McKinney 1998) appears to be the result of a combination of forces, including both extinction and colonization. However, California's native fish fauna has not exhibited large numbers of extinctions (six statewide or 9% of the native fauna) (Moyle 2000), and therefore, extinction does not contribute much to the overall process of biotic homogenization within the state. In contrast, species introductions have significantly altered the state's aquatic fauna. A good example is the Salton Sea. Historically the Salton sink, a low point in the Imperial Valley, would periodically fill with flood water from the Colorado River. Fishes from the river would colonize this habitat, although the created water bodies were generally short-lived (Moyle 2000). In 1905 a large flood diverted the entire flow of the Colorado River into the sink until 1907. Prior to 1905, the only species native to the Salton sink was the desert pupfish (*Cyprinodon macularius*) (Moyle 2000). Since that time, this shallow salty lake (currently known as the Salton Sea), has experienced numerous introductions of alien fish species from a variety of sources (e.g., aquaculture, sport fishing, aquarium escapees, etc.). Currently the lake's fauna is dominated by exotic species, thus giving rise to the lake's spectacular 2500% increase in species richness since 1850.

Table 5. Univariate (correlation) and multivariate (complete and reduced models) relationship of environmental variables to the Jaccard similarity (JFS score) of original to current fish faunas within each California watershed. Large negative coefficients are those associated with greater alteration in species assemblages. Pearson's correlation coefficient significant at P<0.05*, P<0.01**, P<0.001*** (Bonferroni adjusted values).

Variable	Pearson's Correlation Coefficient		Complete Model (F=11.02, Adj. R²=0.72), Std. Coefficient	P	Stepwise Model (F=33.83, Adj. R²=0.82), Std. Coefficient	P
Dams	-0.395	*	-0.222	.113	-0.181	.012
Reservoir area	-0.635	***	0.037	.815		
Ditch density	-0.599	***	0.110	.445		
Aqueduct density	-0.664	***	-0.354	.006	-0.405	<.001
Proportion developed	-0.375	*	-0.019	.881		
Proportion agriculture	-0.321		-0.127	.326		
High protection status	-0.156		0.037	.712		
Mean annual rainfall	0.649	***	0.297	.006	0.333	<.001
Mean elevation	-0.650	***	-0.461	.010	-0.307	<.001
Stream density	0.211		-0.102	.379		
Watershed area	-0.623	***	-0.073	.604		
Dams * Aqueducts					0.187	.012
Rainfall * Elevation					-0.203	.006

The interaction of pattern and scale is extremely important for questions of biodiversity and species invasion (Levin 1992, Rathert et al. 1999). Much of the literature on invasions holds that species introductions may lead to increased alpha (local, or within habitat) diversity, but will tend to decrease both beta (across habitat) and gamma (total, or in this context, statewide) species diversity (Harrison 1993, Meffe and Carroll 1994). However, all of these observations are in turn, scale-dependent. In this study, both alpha (local and/or reservoir level) and gamma (state level) diversity have clearly increased, in some cases many-fold, as a consequence of introductions.

The outcome for beta diversity in this study depends again on the scale of observation. At the broad scale of zoogeographic provinces, similarity among fish faunas has increased, beta diversity has decreased, and homogenization of fish faunas is evident. At a lower scale of watersheds within provinces, similarity has not clearly increased, and in fact has decreased significantly in both the North and South Coast provinces (Figure 3). Both of these regions historically had fairly uniform within-province faunas as the result of many anadromous species and/or Pleistocene connections during periods of low sea level (Moyle 2000). The haphazard nature of introductions within individual watersheds in these provinces is likely the cause of the paradoxical result of decreased similarity (increased beta diversity) since 1850. In contrast, the Central Valley province remains relatively interconnected, and similarities among watersheds have retained much of their original pattern.

How are these patterns related to the process of physical habitat homogenization? Two lines of evidence are relevant here. First, at a given scale of observation (within or across zoogeographic provinces), comparisons among reservoirs show a greater tendency towards increased homogenization than comparisons among all watersheds. Reservoirs are among the most homogeneous of human-created aquatic habitats; our results suggest that their fish faunas have become more homogeneous than that of present-day watersheds in general.

Second, the overall change in fish faunas within California watersheds is significantly related to several measures of anthropogenic habitat alteration. These measures of aquatic habitat alteration, including number of dams, reservoir area, and aqueduct density, all appear related to increased faunal change, as does watershed development in general. The interaction between the variables dams (or reservoir area) and aqueducts suggests that these types of impacts do not so much enhance each other as replace each other. A high level of either factor is sufficient to ensure a large alteration in fish faunas, whatever the value of the other factor. For example, a watershed with many dams will likely have many introductions and extinctions, whether or not it has many aqueducts. Similarly, a watershed with many km of aqueducts will have a highly altered fish fauna even if it has few dams.

Two important natural factors associated with the degree of change in fish faunas are rainfall and elevation. Both factors are significant elements in models predicting native diversity and number of introduced species in California watersheds (T. Light unpublished data). The interaction of rainfall and elevation is complex, but may be related mainly to the vagaries of California geography and settlement patterns. High precipitation watersheds occur mostly in the North Coast (and Klamath) and northern Central Valley zoogeographic provinces, while low precipitation watersheds are concentrated in the Southern California, Great Basin, and Salton Sea provinces. Within the higher precipitation parts of the state, the Central Valley watersheds, which reach up into the Sierra, have higher average elevations, and are also much more influenced by invasions and extinctions, probably mainly due to denser human population and associated water development (Table 4). Within lower rainfall parts of the state, the low-elevation, very hydrologically altered South Coast province is more affected by extinctions and invasions than the higher-elevation, sparsely settled Great Basin. Similarly, looking just at the lower elevation provinces (North Coast and Southern California), low-rainfall Southern California has a much more altered fish fauna than the wetter north.

6. CONCEPTUAL FRAMEWORK OF BIOTIC HOMOGENIZATION

In light of the above analysis and given the perceived link between abiotic change and biotic homogenization, we developed a heuristic framework that relates the processes of habitat alteration/physical homogenization with species introductions and biotic homogenization (Figure 5). Within this framework, homogenization of a region's fish fauna is the outcome of two separate ecological forces, a loss of native species (extinction) and the introduction of new species (colonization). These two forces can operate independently or in concert to increase spatial biotic similarity. This link between forces is extremely important, as species introductions are known to directly influence species extinctions (Taylor et al. 1984, Moyle and Cech 1996, Williamson 1996) as well as influence the physical characteristics of a habitat (Taylor et al. 1984). Such a link between extinction and colonization was suggested by Lodge (1993) when examining aquatic ecosystem responses to climate change.

The physical changes discussed above can create environments that are hostile to native species, favorable to non-native species, or both. In certain aquatic environments habitat alteration is necessary for particular non-native species to invade. For example, the suite of non-native centrarchids (*Pomoxis nigromaculatus, P. annularis, Lepomis macrochirus, L. cyanellus, Micropterus salmoides*,) introduced into Isabella Reservoir (Kern River, 1000m elevation in the Sierra Nevada, Kern Co.) would not likely persist if the reservoir was removed. Mid-elevation snow-melt rivers such as the Kern River are generally too cold and swift to support resident populations of these more lentic species.

The conceptual framework is an attempt to provide structure to the relationships between the pertinent ecological forces. Testing of the various links in this framework has yet to be done, however the present analysis supports the idea that abiotic forces can lead to biotic homogenization and that species introductions are an integral part of the overall equation.

13. *Homogenization of California's Fish Fauna*

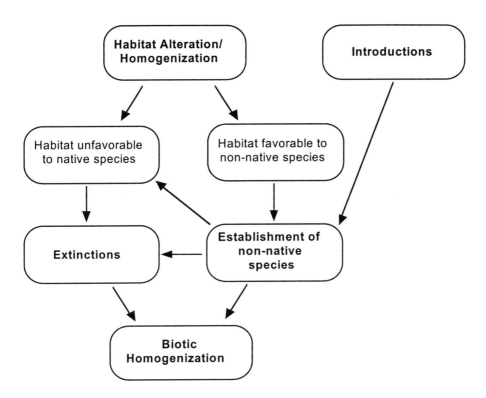

Figure 5. Conceptual framework of the biotic homogenization process including habitat alteration and species introductions.

The results of this study suggest that anthropogenic homogenization of the aquatic environment has contributed to homogenization of the fish fauna in California. This result may not hold in all systems. In some aquatic systems it appears that faunal homogenization (through introduction) is in fact, not facilitated by abiotic or anthropogenic change. For example three species of trout, rainbow, brown and brook, (*Oncorhynchus mykiss, Salmo trutta, Salvelinus fontinalis* respectively) have been introduced to over 50 countries on every continent making them some of the most widely introduced fishes (Lever 1996). Yet, many of the systems into which they have been introduced are not associated with large human populations. Reasons for these taxa's cosmopolitan distribution are related more to environmental requirements (and human affection for this family of sportfishes) than to anthropogenic habitat alterations. The habitat in which these species thrive, is cool, well-oxygenated, gravel-bottomed waters from near sea level up to 4,000 m (Lever 1996, Moyle 2000). This aquatic habitat is extremely common around the globe and not often associated with human disturbance. Instead, the abundance of this habitat is strongly related to previous periods of glacial activity (Moyle 2000), a geographically common and temporally episodic force of extreme abiotic homogenization.

Freshwater faunal assemblages around the world have been dramatically altered in the last century (Williams et al. 1989, Lever 1996), and as we begin a new century, it is important to ask questions regarding the fate of our aquatic heritage. One of these questions should be; do we want rivers, lakes and estuaries to be dominated by a suite of cosmopolitan exotic species? Since 1850, it is clear that the richness of California's fish fauna has significantly increased, yet at some cost to native species. Biodiversity has been variously defined and given different meanings at multiple levels of biological organization (Power et al. 1996). Yet in its largest sense, biodiversity can be seen as an end in itself, whereby the maximal species diversity around the globe should be maintained (Wilson 1989, Power et al. 1996). The goal of preserving biodiversity is clearly at odds with the process of biotic homogenization, or the blending of biological variety that we see in California's fish fauna.

ACKNOWLEDGEMENTS

We would like to thank, T.L. Ford, R.E.Leidy, J. Setka, L. Pardy, R.A. Swenson and F. Kopperdahl for their assistance with the fish data. Thanks also to two anonymous reviewers for their comments. M.P. Marchetti was supported as a post-doctoral researcher by the University of California, Water Resources Center as part of Water Resources Center Project UCAL-WRC-W-880. T. Light was supported by an EPA STAR fellowship.

REFERENCES

Bennett, W.A, P.B. Moyle. 1996. Where have all the fishes gone? Interactive factors producing fish declines in the Sacramento-San Joaquin estuary. In: San Francisco Bay: the ecosystem (Hollibaugh J.T. ed.), Pp: 519-542.

Chamberlin, T.W., R.D. Harr, F.H. Everest. 1991. Timber harvesting, silviculture, and watershed processes. pp. 181-205. In: Influences of forest and rangeland management on salmonid fishes and their habitats (Meehan, W.R. ed.), American Fisheries Society Special Publication 19. Bethesda MD.

Dill, W.A., A.J. Cordone. 1997. History and status of introduced fishes in California, 1871-1996. Fish Bulletin 178. State of California Department of Fish and Game. 414 pp.

Espey, W.H.,. 1969. Urban effects on the unit hydrograph.. In: Effects of watershed changes on stream flow (Moore, W.L., C.W. Morgan, eds.), University of Texas Press, Austin, Pp: 215-228

Furniss, M.J., T.D. Roelofs, S. Yee. 1991. Road construction and maintenance. In: Influences of forest and rangeland management on salmonid fishes and their habitats (Meehan, W.R. ed.), American Fisheries Society Special Publication 19, Bethesda, MD, Pp: 297-323.

Harrison, S. 1993. Species diversity, spatial scale and global change. In: Biotic Interactions and Global Change (P.M. Kareiva, J.G. Kingsolver, and R.B. Huey. eds.), Sinauer Ass. Inc. Mass, Pp: 388-401.

Holtby, L.B. 1988. Effects of logging on stream temperatures in Carnation Creek, British Columbia, and associated impacts on the coho salmon (Oncorhynchus kisutch). Canadian. Journal of Fisheries and Aquatic Science 45:502-515.

Knapp, R.A., K.R. Matthews. 1996. Livestock grazing, golden trout, and streams in the golden trout wilderness, California: Impacts and management implications. North American Journal of Fisheries Management 16:805-820

Lazaro, T.R. 1979. Urban hydrology – a multidisciplinary perspective. Ann Arbor Science Publishers Inc., Ann Arbor, 249pp.

Lever, C. 1996. Naturalized fishes of the world. Academic Press, London.

Levin, S.A. 1992. The problem of patten and scale in ecology. Ecology 73(6):1943-1967.

Lodge, D.M. 1993. Species invasions and deletions: community effects and responses to climate and habitat change. In: Biotic Interactions and Global Change (P.M. Kareiva, J.G. Kingsolver, and R.B. Huey. eds.), Sinauer Ass. Inc. Mass, Pp: 367-387.

Meffe, G.K., and C.R. Carroll. 1994. Principles of Conservation Biology. Sinauer Associates Inc. Sunderland, Massachusetts, 600 pp.

McKinney, M.L. 1998. On predicting biotic homogenization: species-area patterns in marine biota. Global Ecology and Biogeography Letters 7:297-301.

Mount, J.F. 1995. California rivers and streams. University of California Press, Berkeley. 359pp

Moyle, P.B., and J.J. Cech Jr. 1996. Fishes: An Introduction to Ichthyology. 3rd ed. Prentice Hall. New Jersey.

Moyle, P.B., T. Light. 1996. Fish invasions in California: Do abiotic factors determine success? Ecology 77(6):1666-1670.

Moyle, P.B. In press. Inland fishes of California, 2nd edition. University of California Press, Berkeley.

Moyle, P.B. and J.E. Williams. 1990. Biodiversity loss in the temperate zone: decline of the native fish fauna of California. Conservation Biology 4:275-284

Onorato, D., K.R. Marion, R.A. Angus. 1998. Longitudinal variations in the ichthyofaunal assemblages of the Upper Chaba River: Possible effects of urbanization in a watershed. Journal of Freshwater Ecology 13(2):139-154.

Petts, G.E. 1984. Long-term consequences of upstream impoundment. Environmental Conservation 7:325-332

Platts, W.S. 1991. Livestock grazing. In: Influences of forest and rangeland management on salmonid fishes and their habitats (Meehan, W.R. ed.), American Fisheries Society Special Publication 19. Bethesda, MD, Pp: 389-423.

Poff N.L., J.D. Allan. 1995. Functional organization of stream fish assemblages in relation to hydrological variability. Ecology 76(2):606-627.

Power, M.E., W.E., Dietrich, J.C. Finlay. 1996. Dams and downstream aquatic biodiversity: Potential food web consequences of hydrologic and geomorphic change. Environmental Mananagement 20(6):887-895.

Rathert, D., D. White, J.C. Sifneos, and R.M. Hughes. 1999. Environmental correlates of species richness for native freshwater fish in Oregon, USA. Journal of Biogeography 26:257-273.

Reisner, M. 1986. Cadillac Desert: The American West and its disappearing water. Viking Penguin, New York.

Richards, C., L.B. Johnson, G.E. Host. 1996. Landscape-scale influences on stream habitats and biota. Canadian Journal of Fisheries and Aquatic Sciences 53(Suppl. 1):295-311.

Scott, J.B., C.R. Steward, Q.J. Stober. 1986. Effects of urban development on fish population dynamics in Kelsey Creek, Washington. Transactions of the American Fisheries Society 115:555-567.

Taylor, J.N., W.R. Courtenay, Jr., and J.A. McCann. 1984. Known impacts of exotic fishes in the continental United States. In: Distribution, Biology, and Management of Exotic Fishes (W.R. Courtenay Jr. And J.R. Stauffer Jr. eds.), John Hopkins University Press, Baltimore, MD, Pp: 322-373.

Weaver, L.A., G.C. Garman. 1994. Urbanization of a watershed and historical changes in a stream fish assemblage. Transactions of the American Fisheries Society 123:162-172

Williams, J.E., J.E. Johnson, D.A. Hendrickson, S. Contreras-Balderas, J.D. Williams, M. Navarro-Mendoza, D.E. McAllister and J.E. Deacon. 1989. Fishes of North America endangered, threatened or of special concern: 1989. Fisheries 14(6):2-20.

Williamson, M. 1996. Biological invasions. Chapman and Hall. London.

Wilson, E.O. (ed), 1988. Biodiversity. National Academy Press. Washington D.C.

Yoshiyama, R.M., F.W. Fisher, P.B. Moyle. 1998. Historical abundance and decline of chinook salmon in the central valley region of California. North American Journal of Fisheries Management 18:487-521.

Contributors

Trip Armstrong • Department of Wildlife, Fish, and Conservation Biology, University of California, Davis, California, USA

Jonathan E.M. Baillie • Department of Biology, Imperial College, Silwood Park and Institute of Zoology, Zoological Society of London, London, UK

Peter M. Bennett • Institute of Zoology, Zoological Society of London, Regent's Park, London, UK

Robert B. Blair • Department of Zoology, Miami University, Oxford, Ohio, USA

Thomas Brooks • Center for Advanced Technologies and Department of Biological Sciences, University of Arkansas, Fayetteville, Arkansas, USA

Debbie A. Carino • Department of Botany, University of Hawaii, Honolulu, Hawaii, USA

Rob Channell • Department of Biological Sciences, Fort Hayes State University, Hays, Kansas, USA

Curtis C. Daehler • Department of Botany, University of Hawaii, Honolulu, Hawaii, USA

Jeffrey R. Duncan • Department of Ecology and Evolutionary Biology and Water Resources Research Center, University of Tennessee, Knoxville, Tennessee, USA

Joaquin Feliciano • Department of Wildlife, Fish, and Conservation Biology, University of California, Davis, California, USA

David Gammon • Department of Zoology, Brigham Young University, Provo, Utah, USA

Kevin J. Gaston • Biodiversity and Macroecology Group, Department of Animal and Plant Sciences, University of Sheffield, Sheffield, UK

Jessica L. Green • Department of Nuclear Engineering, University of California, Berkeley, California, USA

John Harte • Energy and Resources Group and Department of Environmental Science, Policy and Management, University of California, Berkeley, California, USA

Zeb Hogan • Department of Wildlife, Fish, and Conservation Biology, University of California, Davis, California, USA

Jeffrey S. Kauffman • Department of Biology, Indiana University, Bloomington, Indiana, USA.

Melanie Kershaw • The Wildfowl and Wetlands Trust, Slimbridge, Gloucestershire, UK

Theo Light • Department of Wildlife, Fish, and Conservation Biology, University of California, Davis, California, USA

Eric T. Linder • Department of Zoology, Brigham Young University, Provo, Utah, USA

Julie L. Lockwood • Department of Environmental Studies, University of California, Santa Cruz, California, USA

Mark V. Lomolino • Oklahoma Biological Survey, Oklahoma Natural Heritage Inventory and Department of Zoology, University of Oklahoma, Norman, Oklahoma, USA

Michael P. Marchetti • Department of Wildlife, Fish, and Conservation Biology, University of California, Davis, California, USA

Brian A. Maurer • Department of Fisheries and Wildlife and Department of Geography, Michigan State University, East Lansing, Michigan, USA

Michael L. McKinney • Department of Geology and Department of Ecology and Evolutionary Biology, University of Tennessee, Knoxville, Tennessee, USA

Peter B. Moyle • Department of Wildlife, Fish, and Conservation Biology, University of California, Davis, California, USA

Ian P.F. Owens • Department of Zoology and Entomology, University of Queensland, Brisbane, AUSTRALIA

A. Ostling • Energy and Resource Group, University of California, Berkeley, California, USA

David R. Perault • Department of Biology and Environmental Science, Lynchburg College, Lynchburg, Virginia, USA

Kaustuv Roy • Department of Biology, University of California–San Diego, La Jolla, California, USA

Daniel Simberloff • Department of Ecology and Evolutionary Biology, University of Tennessee, Knoxville, Tennessee, USA

Gregory A. Smith • Department of Biology and Oklahoma Biological Survey, Norman, Oklahoma, USA

Diego P. Vazquez • Department of Ecology and Evolutionary Biology, University of Tennessee, Knoxville, Tennessee, USA

Joshua Viers • Information Center for the Environment, University of California, Davis, California, USA

Thomas J. Webb • Biodiversity and Macroecology Group, Department of Animal and Plant Sciences, University of Sheffield, Sheffield, UK

INDEX

Abbot's booby, 71
Acari, 110
Accipter gentilis, 132
Acrocephalus, 69
    Acrocephalus seychellensis, 71
    Acrocephalus paludicola, 71
Aelothripidae, 113
Africa, 103, 182
Alabama, 250
Albatrosses, 69, 71
Alectoris chucar, 127
Alectoris graeca, 127
Aleyrodidae, 110
Alien species, 34, 81-87, 90-91, 93, 97
Allee effect, 167
Allelopathy, 83
Allozyme analyses, 86
Ambystoma tigrinium, 230
Ammoperdix heyi, 127
Antartica, 114
Anatidae, 11, 131, 205
Anseriformes, 59-64, 73
Anthocoridae, 109-111
Antillean euphonia, 127-128
Aphids, 110, 114

Aphididae, 110, 114
Apis mellifera, 97
Apterigidae, 214
Apteryx owenii, 132
Aquatic warbler, 71
Aquila chrysaetes, 230
Arrenophagus albitibae, 111
Asteraceae, 87, 94
Athrenus museorum, 114-115
Atlantic Ocean, 203
Atlantic smooth cordgrass, 86
Australia, 6, 127, 224-225

Bald Ibis, 209
Barndoor skate, 73
Bidens, 96
Biodiversty, 1-2
Biogeographers, 158
Biological invasions, 19-20, 25-29, 33-34
Birdlife International, 125-127
Black-footed ferret, 230
Black-tailed deer, 233
Black-tailed prairie dog, 230
Blue-headed euphonia, 127
Blue-winged warbler, 13

Brazil, 11, 131
Britain, 7, 14, 28, 180
British Isles, 87, 96
Brotogeris chiriri, 127
Brotogeris versicolorus, 127
Brown quail, 127
Bubo virginianus, 132
Buteo regalis, 230
Burrowing owl, 230

Cacatua gulerita, 126
California, 8, 21, 23, 34-40, 45-48, 51-54, 127, 237, 239, 253, 259-260, 262-263, 265, 271, 273-276
California condor, 132, 209
California cordgrass, 86
Caliroa ceradi, 112
Canada, 85, 125
Canadian hawthorn, 82, 84
Canary-winged parakeet, 127
Carassius auratus, 255
Carpobrotus chilensis, 85
Carpobrotus edulus, 85
Carpodacus mexicanus, 160, 171
Casuariidae, 214
Celastrina laden, 44
Cenezoic, 20
Centaurea diffusa, 111
Centaurea maculosa, 111
Centrarchidae, 263
Charadrius montanus, 230
Chatham Island snipe, 132
Chimney swift, 51
Christmas Bird Count (CBC) 164
Ciconiiformes, 217
Clethrionomys gapperi, 229
Cnephasia longana, 113
Coast redwoods, 47
Coccidae, 110
Coccinellidae, 109-111
Coenocorypha pussila, 132

Colias eurytheme, 44
Collocolia fuciphaga, 127
Collocolia inexpectata, 127
Columbidae, 11, 131, 205, 211, 214, 217
Conservation biology, 125, 229
Continental ecosystems, 3-4
Coturnix australis, 127
Coturnix ypsilophora, 127
Crataegus douglassi, 87
Crataegus monogyna, 83-84, 87
Crataegus punctata, 82-84
Crimson-backed tanager, 126
Common deer mouse, 233
Costa Rica, 58
Crotalus viridis, 230
Cucilidae, 210
Cygnus baccinator, 72
Cynipidae, 114
Cyprinidae, 247, 255
Cyprinodon macularius, 271
Cynomys ludovicianus, 226
Cytisus scopanus, 114

Danaus plexippus, 44
Darwin, Charles, 233
Desert pupfish, 271
Diomedea amsterdamensis, 71
Diomedea albatrus, 71
Diomedea exulans, 71
Diprotodons, 225
Diptera, 114
Drepanididae, 201

Easter Island, 204, 229
Eastern hemlock, 47
Ecological Flora Database, 104
Ecological specialist, 13
Ectopistes migratorius, 58
Edible-nest swiftlet, 127
Ediths checkerspot butterfly, 179
Elaphus maximus, 237
Encrytidae, 111

Endagered Species Act, 34
Endemic-area relationship (EAR), 182-184, 187-188, 190-191, 193, 197
Epilobium, 87
Epimecis detexta, 111
Eremophila alpestris, 228
Eulophidae, 111
Euphonia musica, 127
Euphonia musica sensu stricto, 128
Euphorbia, 87
Euphydryas editha, 179
Eurasia, 226
European hawthorn, 83-84, 87
European honeybees, 97
European purple loosestrife, 85
European Nations, 5-6
European starling, 159, 164, 166-167
Eurytomidae, 113
Extinction-biasing traits, 11
Extinction-prone, 11, 57, 201
Extinction risk, 57-59, 61, 63, 206-213, 217
Extralimital, 23, 25-26

Ferruginous hawk, 230
Fiji, 127
Forest deer mouse, 233
Fringillidae, 214
Fundulus albolineatus, 246
Furnariidae, 214

Galapagos Islands, 85
Gallinula comeri, 128
Gallinula nesiotis, 128, 132
Gasterophylus, 114
Geopsittacus occidentalis, 209
Germanium, 87
Geronticus eremita, 209
Glaucomys sabrinus, 233
Global warming, 181, 184

Gnypetoscinus queenslandiae, 239
Golden-winged warbler, 13
Golden eagle, 230
Goldfish, 255
Gondwanaland, 217
Goshawk, 132
Gossypium barbadense, 94
Gossypium darwinii, 85
Gossypium tomentosum, 94
Grasshopper mouse, 230
Great Horned Owl, 132
Griffon vulture, 132
Gruidae, 213
Gryllptalpidae, 112
Gymnogyps californicus, 132, 209
Gyps fulvus, 132

Habitat degradation, 34
Harelip sucker, 250
Hawaii, 4, 82, 91, 94, 96, 111, 125, 127, 204-205
Hedyotis, 96
Hesperioidea, 36
Heterarthrus nemoratus, 112
Heterogeneity, 1
 ecological heterogeneity, 159
Heteroptera, 110
Hippolais, 69
Holocene, 27, 30, 203-204, 210, 225, 228
Homogenization, 1, 5, 13-14, 20-21, 29, 34, 37, 45-47, 51, 116, 157, 158, 163, 174, 239, 254, 271, 275
 abiotic homogenization, 179, 259, 262, 276
 anthropogenic homogenization, 224-225
 biotic homogenization, 1-2, 14-15, 19-20, 26, 30, 160,

Homogenization (*cont'd*)
    162, 245, 259, 262, 271, 274
  ecological homogenization, 13
  faunal homogenization, 275
  genetic homogenization, 81, 84
  geographic homogenization, 125
  global biotic homogenization, 198
  habitat homogenization, 2
  homogeneity, 44
  spatial homogenization, 10, 13, 248
  taxonomic homogenization, 12-13, 246
  temerature homogenization, 181
Hoplocampa brevis, 112
Horned Lark, 230
House Finch, 37, 159, 171
Hybridization, 12-13, 46, 81-87, 90-91, 94, 96-98
  native-alien hybridization, 87, 90-91, 94, 96-97
Hymenoptera, 109, 111-114, 116
Hypoderma, 114
Hypsipetes borbonmicus, 128
Hypsipetes olivaceus, 128

Incas, 229
Indian Ocean, 203-205
Introgressants, 84-85, 99
IUCN, 201-202, 212

Jamaica, 94
Japan, 86

Lagochila Lacera, 250
Laniidae, 214
Lantana camara, 85
Lantana depressa, 85

Lavinia symmetricus, 239
Lawton, John, 201-202
Laysan finch, 127, 132
Lepomis cyanellus, 274
Lepomis macrochirusi, 274
Lipochaeta, 94
Little spotted kiwi, 132
Local invaders, 37, 46, 52-53
Loxodonta africana, 237
Lythrum alatum, 85
Lythrum salicaria, 85

Madagascar, 204
Mammalia, 225
Mammuthus primigenius, 226
Maui, 94
Mauritius, 205
Mauritius Bulbul, 128
Mediterranean Sea, 237
Megapodiidae, 214
Megisto cymele, 53
Menuridae, 214
Mesites, 218
Mexico, 4, 6, 110
Micropterus salmoides, 274
Mississippi, 254
Montane shrew, 233
Morone saxatilis, 263
Morus alba, 85
Morus rubra, 85
Mountain plover, 230
Mustela nigripes, 230

Nature conservancy, 34
  Nature Conservancy Natural Heritage Network, 4
New Zealand, 127, 133, 204-205, 225
Nihoa finch, 127, 132
North America, 103, 117, 159-160, 164, 166-167, 171, 179, 225, 229-230, 232, 239, 246, 250

Index

North American Breedind Bird Survey (BBS), 160-161, 164-166, 171
North American Non-Indigenous Arthropod Database (NANIAD), 111
Nyctanus violacea, 132

Odocoileus hemionus, 233
Ohio, 8, 34-40, 45-48, 51, 53-54
Oestridae, 114
Olivaceus Bulbul, 128
Oncorhynchus mykiss, 255, 275
Onychomys leucogaster, 230
Oregon, 87
Orthonychidae, 214
Orthoptera, 112
Oryza rufipogon, 85
Oryza sativa, 85
Osterus, 114

Pacific islands, 203, 205
Pacific Ocean, 260
Paleozoic, 1-2, 19
Palm trees, 47
Panama, 58
  Isthmus of Panama, 19
Pangea, 19
Papasula abboti, 70
Paplinoidea, 36
Pardalotidae, 214
Parisitoids, 109, 113, 116-117
Parrot, 9, 210-211, 214, 217
Passenger pigeon, 58
Percidae, 247
Peromyscus maniculatus, 233
Peromyscus oreas, 233
Phasianidae, 11, 131, 210-211, 214, 217
Phrynosoma cornutum, 230
Phylogenetic patterns of rarity, 57-58
Picidae, 210-211

Picea, 114
Pieris rapae, 44
Pin oak, 47
Pleistocene, 19-23, 25-30, 224, 237
Pliocene, 19
Poaceae, 87
Pomoxis annularis, 274
Pomoxis nigromaculatus, 274
Polynesian rat, 204
Portulaca lutea, 94
Portulaca oleracea, 94
Prairie rattlesnake, 230
Prickly forest skink, 239
Pristophora erichsonni, 113
Procellariidae, 211, 214
Psittacidae, 11, 131, 210-211, 214, 217
Psyllids, 110
  Psyllidae, 110
Pulicidae, 114

Rainbow trout, 255
Raja laevis, 73
Rallidae, 205, 211, 214, 217
Ramphocelus dimidiatus, 126-127
Ramphocelus passerinii, 127
Rattus exulans, 204
Red-backed voles, 233
Red List of Threatened Species, 127, 132
Red mulberry, 85
Reunion, 205
ROBO Database, 113
Rodrigues, 205
Rosa, 90
Rosacae, 87
Rubus hawaiensis, 94
Rubus rosifolius, 94
Rumex, 87

Salmo trutta, 275

Salmonidae, 263
Salvensus fontinalis, 275
Sand partridge, 127
Scaphiopus, 230
Scarlet-rumped Tanager, 127
Scotch Broom, 113-114
Seychelles warbler, 71-72
Shannon diversity, 34, 47
Short-tailed albatross, 71
Siphonaptera, 114
Sorex monticolus, 233
Sorex trowbridgii, 233
South Africa, 4, 6, 125
South America, 112, 179
Southern flying squirrel, 233
Spadefoot toad, 230
Spartina, 13
  Spartina alterniflora, 86
  Spartina foliosa, 86
Species-area curve, 2, 9
Species-area effects, 9
Species-area relationship (SAR), 181-184, 190, 193, 196-198
Species diversity, 9
Species richness, 14, 34, 47, 58, 181, 193-194, 196, 215, 217, 251, 255, 263
Speciation, 63, 66-68, 73, 215
  allopatric speciation, 65, 67
Speotyo cunicularia, 230
Spermophilus tridecemlineautus, 230
Spheniscidae, 214
Spilogale putorius, 233
Spotted skunk, 233
Stegobium paniceum, 115
Stizostedion vitreum, 255
Striped bass, 263
Sturnus vulgaris, 159, 164
Sulphur-crested cockatoo, 126
Swamp quail, 127
Swift fox, 230

Tahiti, 127
Taiwan, 85
Tasmania, 127
Tanzania, 6
Taraxacum officinale, 85-86
Taraxacum platycarpus, 85-86
Taxonomic selectivity, 26
Telespiza cantans, 127, 132
Telespiza ultima, 127, 132
Tennessee, 246-250, 252-255
Tennessee Valley Authority, 247
Tennessee Department of Environment and Conservation (TDEC), 248
Tenthredinidae, 112
Tephritidae, 114
Texas horned lizard, 230
Thirteen-lined ground squirrel, 230
Thripidae, 113
  Thrips tabaci, 113
Thysanoptera, 113
Tichospius distraeae, 111
Tiger Salamander, 230
Trifolium, 87
Tristan moorhen, 128
Trochilidae, 214
Trogonidae, 214
Trowbrodge shrew, 233
Trumpeter swan, 72
Turnicidae, 214

United States, 4, 6, 8, 2, 34, 85, 104-106, 108, 111-115
United Sates Army Corps of Engineers, 247
US Biological Survey, 230
US Geological Survey, Biological Resources Division, 161
Urban gradient, 8-9, 34-35, 38-39, 46

Urbanization, 33-38, 40, 45-47, 51-54, 163, 262
Urophora, 111, 114

Verbascum, 87
Vulpes velox, 230

Wandering albatross, 71
Wedelia trilobata, 94
White-line topminnow, 250

White-throated swift, 51
White mulberry, 85
Wildfowl and Wetland Trust, 60
Wooly mammoth, 226, 236

Yellow-chevroned parakeet, 127
Yellow-crowned night heron, 132

Zebra mussel, 27, 254
Zosteropidae, 214